WHOLES, SUMS AND UNITIES

PHILOSOPHICAL STUDIES SERIES

VOLUME 97

The titles published in this series are listed at the end of this volume.

WHOLES, SUMS AND UNITIES

by

ARIEL MEIRAV
University of Haifa,
Israel

KLUWER ACADEMIC PUBLISHERS
DORDRECHT / BOSTON / LONDON

A C.I.P. Catalogue record for this book is available from the Library of Congress.

ISBN 978-90-481-6442-4

Published by Kluwer Academic Publishers,
P.O. Box 17, 3300 AA Dordrecht, The Netherlands.

Sold and distributed in North, Central and South America
by Kluwer Academic Publishers,
101 Philip Drive, Norwell, MA 02061, U.S.A.

In all other countries, sold and distributed
by Kluwer Academic Publishers,
P.O. Box 322, 3300 AH Dordrecht, The Netherlands.

Photograph pp. 23–24:
Giuseppe Arcimboldo, *L'Ortolano.* © *Museo Civico Ala Ponzone – Pinacoteca, Cremona, Italy.*

Printed on acid-free paper

Printed in the Netherlands.

Table of Contents

Part Three: Unities

Preface

The well-worn claim that a whole is greater than the sum of its parts means many things to many people. Nevertheless, it may be thought to express a particular widespread intuition about the nature of wholes. It indicates, however inadequately and vaguely, that wholes are in some way surprizing, different to what might be expected, non-trivial in their characteristics. Perhaps it hints at the impression that if a whole is composed of several parts, then to tell us all we may wish to know about each of these parts is still not to tell us something crucial about the whole itself. This book may be viewed as an attempt to justify and develop this intuition by focusing on what I take to be a neglected dimension in the nature of wholes. To attend to this dimension is to consider different *ways* in which parts, irrespective of their individual characters and of the relations between them, can compose a whole. My contention will be that contemporary metaphysical discussion by and large fails to notice this dimension, and that this may explain some of its more serious difficulties in treating questions about wholes and parts. I shall suggest that a study of this dimension may have far reaching philosophical consequences.

The task of exploring this neglected dimension is reflected in the structure of the book as follows. In Part One I aim to show that the theoretical space within which a theory of wholes and parts might be devised is larger than is often assumed. This prepares us for the possibility of rejecting certain widely accepted assumptions about wholes. In Part Two I isolate assumptions shared by contemporary theories, showing how they stand in the way of considering the dimension I am referring to, and arguing that they lead to decisive theoretical weaknesses. Finally, in Part Three, I develop a theory - the theory of Unities - which rejects these assumptions, and demonstrate the explanatory power of this theory. Before proceeding to attend to these tasks, however, I devote an introductory chapter to explaining the significance of the project, the motivation for it, and the context in which it is undertaken.

The book is based on and developed from my doctoral dissertation at the University of Edinburgh (1997). I wish to thank all those who have helped, inspired and criticised my work in this field both during the Edinburgh years and later as a member of the departments of philosophy at Tel-Aviv University and the University of Haifa. I would like to mention in particular Leora Bilsky, Pete Campbell, William Charlton, Paolo Crivelli, Fritz-Gregor Herrmann, Hagi Kenaan, Vered Lev Kenaan, E. J. Lowe, Ilse Mogensen, Gregory Owcarz, Jennifer Richards, David Robinson, Theodore Scaltsas, Eva Schiedrum, Traute Schiedrum, Barry Smith, Paul Suttie, Christopher Strachan, David Vila Domini, and Ruth Weintraub. I am greatly appreciative of the suggestions made by Kluwer Academic Publishers' anonymous reader in response to a draft of my manuscript. And finally, I would like to express a very special gratitude to Sharon Krishek for her unflinching, loving and wise support.

Tel-Aviv, May 2003

Chapter 1
Introduction

I - Plato's Challenge

Whole and part are among the most pervasive notions that occur in our thinking, and philosophical thought in particular seems to be bound up with them in its very essence. On the face of it, almost anything we care to think about may be considered as a whole and as having parts, or as being itself a part of some greater whole. In attempting to understand ourselves and our environment, we divide or analyze (a whole into parts), we collect or synthesize (parts into a whole), we discern many parts composing one whole, and one whole composed of many parts. When, aspiring to understand, we reflect on the things we see, hear, imagine or conceive, on everyday objects, on sentences or stories or melodies, on historical events, on our own actions, on mathematical structures, on the products of art, on space and on time, on reality considered metaphysically or theologically – we are almost irresistibly led to consider them in the light of the notions of whole and part.

Anything in which we can discern parts is susceptible to two apparently distinct perspectives – a perspective concerned with one thing, the whole, and a perspective concerned with many things, the parts. We may call these the singular and plural perspectives, respectively. The ability to shift from one to the other is something we usually take for granted in thought and communication, so much so that we often do not even notice that our perspective thus undergoes a shift. It is an important ability, for it provides us with the rudiments of an almost universally applicable method of investigation, one which we can hardly imagine doing without. By shifting attention to the parts of something, and studying those parts, it seems that our understanding of the thing itself is increased. It is doubtful whether there can be a science or an explanatory theoretical activity of any sort which does not make extensive use of this vaguely stated methodological principle.

Therefore, the notions of whole and part are philosophically interesting not only because they are widely applicable – that is, because almost everything seems to be either a whole or a part or both - but more importantly because they seem to play a crucial role in theoretical activity in general, the role associated with providing an understanding of a whole with reference to its parts, as well as (though perhaps to a lesser degree) an understanding of parts with reference to the whole which they compose. Our conception of these notions, of the relation between something that is a whole and those things which are its parts, determines our view of the implications which knowledge about parts has for knowledge about the whole. In other words, it determines our interpretation of the results of any investigation into the nature of something which proceeds by considering its parts.

Much hangs, therefore, on our conception of the relations between a whole and its parts. This realization is troubling, and philosophically motivating, in view of the fact that different conceptions of these relations vie for acceptance, and even more so in view of the fact that none of the familiar competing conceptions seems entirely satisfactory. The profound observation that different conceptions are possible was apparently first made by Plato. In a famous passage of the *Theaetetus* he considers the general nature of complexes, where a complex (*sullabē*) is an entity which we take somehow to be made up or to arise out of simple entities described as its elements (Plato clearly thinks that exploring the notion of a complex is relevant to our attempts to answer the question 'What is knowledge?', to which the *Theaetetus* is devoted).[1] A paradigmatic example of a complex and its elements is provided by the case of a syllable and the letters associated with it, respectively – say the syllable SO and the letters S and O. Plato suggests that two contrasting conceptions of the complex are available – as something *identical* to the elements, or as something *distinct* from them. That is to say, what we refer to by 'SO' is either identical to, or distinct from, what we refer to by 'S and O'. It is important not to confuse this with the suggestion that the complex is either identical to or distinct from *each* of the elements - that what we refer to by 'SO' is either identical to, or distinct from, what we refer to by 'S', and what we refer to by 'O'. The claim that SO is identical to each of the letters S and O would seem to be patently absurd, for it implies that S and O are identical to one another. What Plato is suggesting may be perplexing or even plainly mistaken, but it is certainly not a non-starter of this sort.

It turns out that when we adopt the singular perspective (according to which we describe what we see as the complex SO rather than the

---

[1] See *Theaetetus* 202e-205e, and particularly 203c-205a.

elements S and O), we may intend two different sorts of entity, that is, either a complex identical to its elements, or a complex distinct from its elements. And so what at first glance might seem to be a two-tiered conceptual structure resolves itself, or perhaps broadens itself out, into a three-tiered one. Instead of merely considering elements and a complex, we should consider elements (many things), an identical complex (one thing), and a distinct complex (a further one thing). And this raises the question, whether SO the syllable is a complex of the first or second sort. The point is illustrated by the following diagram:

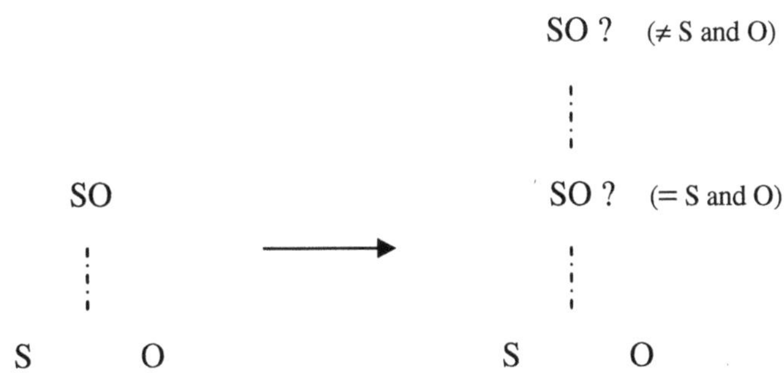

Figure 1

In the same passage Plato reserves the term 'whole' (i.e. '*holon*') for the first sort of complex, the identical one (thus, the passage leaves open the question whether the *complex* is identical to or distinct from its elements, but argues that a *whole* cannot be distinct from its parts and must be identical to them).[2] If we ignore this terminological decision, however, we may derive from his discussion the suggestion that two distinct conceptions of a whole are available, in principle, one according to which the whole is identical to its parts, and the other according to which it is distinct from them. And we are challenged to contemplate what might be described as three conceptual levels, that of parts, that of a whole which is identical to them, and that of a whole which is distinct from

---

[2] At 204a Socrates states that 'when a thing has parts, the whole is necessarily all the parts', and by the end of 205a he concludes a proof of this statement.

them. For convenience we may speak of them as the levels of parts, of a 'lower-grade' whole, and of a 'higher-grade' whole, respectively.

If we are concerned about our universally applied theoretical method of learning about wholes by investigating their parts, Plato's suggestion seems to pose an immediate problem. For such a method seems to presuppose certain logical or conceptual relations between facts about parts and facts about the wholes of which they are parts. These, however, must surely depend on what we take the relation between the whole and the parts to be. Even if our inferences based on the investigation of parts are justifiable on the assumption that one of the two conceptions is correct, we now face the problem of justifying that conception, explaining why it rather than the other one should be embraced. We may be systematically mistaken in our theoretical proceedings, if we presuppose a mistaken conception of wholes. It might be, for example, that our theoretical method is based on the assumption that a whole is identical to its parts, and that our theoretical conclusions follow from this assumption, but that they are false or at least entirely unsubstantiated since the assumption is false. Thus Plato's suggestion prompts us to begin to take seriously the need to engage in a careful study of the notions of whole and part.

The problem which Plato poses for us is much larger and more disturbing than that of needing to choose between two allegedly viable conceptions. For it is a striking feature of his account that it seems to give us a clear answer to the question, 'what is a whole?', only in connection with the first conception, that is, in connection with the conception of lower-grade wholes. The first conception tells us that the whole is, simply, all of the parts. Plausible or not, this is at least a clear answer. Regarding the second conception, however, he only tells us that the whole is 'some single form, produced out of them [i.e. the letters], having its own single nature – something different from the letters',[3] which seems to be not much more than saying that it is some one thing which is distinct from the parts. We seem in this case to be told what the whole is not, whereas we want to know what it is. Of course we also know that the whole is something which *has* these parts (and we may know all sorts of things about the parts and the relations between them), but this is also true of the whole according to the first conception. Thus, if a whole is different from its parts, the mere assumption that it has those parts does not help us

---

[3] See 203e.

to determine what it is like, in what *way* it is different from the parts. And so, if we accept the assumption that the syllable SO is not the same as the letters S and O, we seem to be left in the dark as to what it might be. Our hopes of clarifying its nature on the basis of a prior understanding as to what the letters are seem to be thoroughly dashed. Therefore, if we wish to reject the first conception of wholes we are left with the second one which renders them profoundly mysterious.

And there do seem to be good reasons, if not for rejecting the first conception, then at least for viewing it as failing to provide the basis for a complete account of wholes. The conception involves both logical and metaphysical difficulties. The logical objection, voiced by contemporary authors such as David Lewis, Peter van Inwagen, and Peter Simons, among others, targets the implied assumption that identity of one thing to many things makes sense; identity is naturally viewed as a one-one relation, not a one-many relation, and if this is correct one-many identity does *not* make sense.[4] We even seem undecided as to whether to say that the whole and the parts are the same *thing* or the same *things*. If we say that the parts *are* the whole, the objectors urge, we should be interpreted as meaning only that the parts *compose* or *constitute* the whole, which may imply that the 'sum' of the parts is identical to the whole, but not that the parts themselves are identical to the whole. As for the metaphysical objection, it concerns the apparent implication of the first conception that the mere existence of the parts guarantees the existence of the whole, something which we do not generally believe to be true. This point was certainly noticed by Aristotle, and arguably by Plato as well.[5] Clearly one may write S and O on a page without thereby writing SO on that page. Thus even if the logical objection can be answered, the metaphysical objection may leave us unsatisfied with the first conception as a generally valid conception of wholes.

Plato's brief discussion in the *Theaetetus* therefore bequeathed to posterity a challenge. An understanding of wholes is a central requirement of our theoretical lives, and yet wholes are elusive and mysterious.

---

[4] See Doepke 1982, 46-7; Simons 1987, 212; van Inwagen 1994; Lewis 1991 considers the relation between the one and the many to be 'like' identity (p.82), but not quite the same as identity (p.87).

[5] See Aristotle's *Metaphysics* 1041b11-31. In Plato's *Phaedo* 85e-86d Simmias suggests that the soul is something which exists only when parts of the body are appropriately related to one another.

Intuitively it seems that knowledge about the parts of a whole is relevant to knowledge about the whole, and yet, if we are unclear about the relation between the whole and its parts we must be unclear also about the *way* in which the former knowledge is relevant to the latter.

II - Two Approaches to Wholes and Parts

A response to Plato's challenge was soon to come. Aristotle's theory of substance seems to agree with Plato's suggestion that fundamentally different conceptions of a whole are available, while at the same time avoiding the mystery apparently entailed by this suggestion. Echoing Plato's distinction between what I have called lower-grade and higher-grade wholes, Aristotle distinguishes between wholes that are merely 'heaps' ('*soroi*') and wholes that are *bona fide* 'substances' ('*ousiai*').[6] A substance, a living organism for example, is a whole which is in an important sense prior to the organs which are its parts. It is not merely that the life of each organ would whither away if the causal connections between it and the other organs were to be severed. Rather, the organs *depend ontologically* on one another: Each of them requires the others as essential aspects of its own existence. It is a part of what it is for such a part to exist, that the other parts exist and are appropriately related to it causally and spatially. The very identity of each part depends on its playing the role it does as a part of the organism. A hand severed from its owner is a hand by name only.[7] A heap, say a pile of firewood, by contrast, is a whole whose parts do not exhibit such mutual dependence. One implication of this is that the study of substances cannot be approached by investigating the parts of the substance and drawing direct inferences concerning the whole, since an adequate understanding of the parts themselves depends on a prior understanding of the whole and the role they play in its functioning. Note that Plato's account seems to have a similar implication (pending a more positive determination of the

---

[6]See *Metaphysics*, especially book Z Chs.10,17; book H Ch. 6.

[7]See 1035b23-25.

difference between higher-grade wholes and their parts). For it implies that some wholes (if there are wholes of the higher-grade sort) cannot be adequately studied merely by considering their parts and drawing direct inferences regarding the wholes, for the reason that we know that the whole is distinct from the parts, but do not know in what way it differs from them.

In order to see how Aristotle's theory might be viewed as providing an adequate solution to the problem presented by Plato's account (though I will soon want to suggest that it should not be so viewed), we need to be clear about a certain ambiguity in the latter. I have said that it proposes that we think of wholes in terms of a three-tiered structure instead of a two-tiered one. This claim may itself be understood in one of two ways. On the one hand, we might take it as suggesting that whenever some particular parts compose a whole, they compose *either* a lower-grade whole *or* a higher-grade whole – never both. On the other hand, we might take it as suggesting that in some cases parts can compose at the same time both a lower-grade and a higher-grade whole. The two interpretations are suggested by Figures 2a and 2b, respectively. Parts are represented by squares, a lower-grade whole by a divided rectangle, and a higher-grade whole by an undivided one. The arrows represent the relation of parthood.

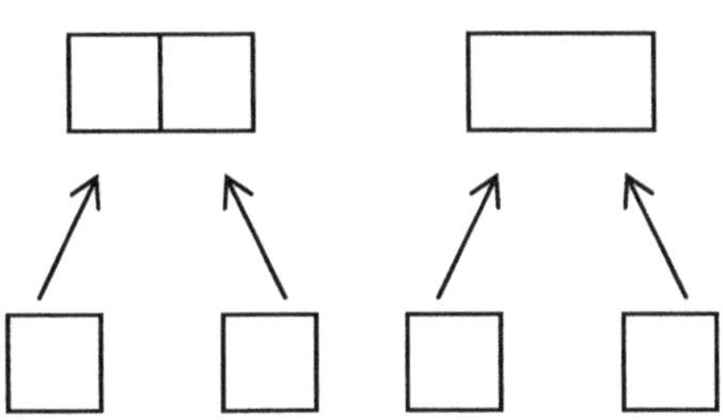

Figure 2a

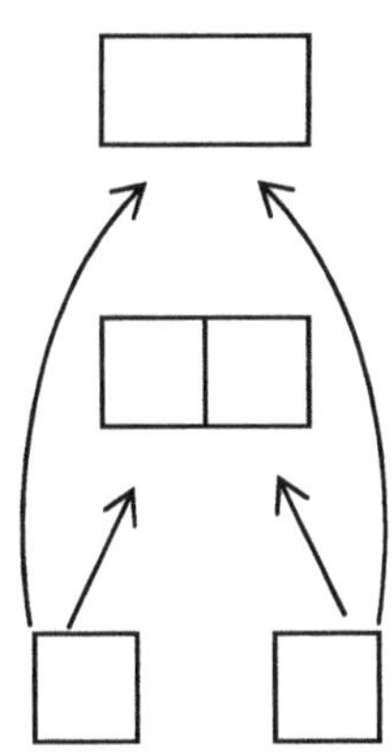

Figure 2b

On the first reading, the case of the syllable SO either involves a lower-grade whole or a higher-grade one, but not both. On the second reading, the case may involve both a lower-grade and a higher-grade whole, and if so we need to decide whether the syllable, strictly speaking, is to be identified with the former or the latter. The second interpretation takes Plato to be suggesting a three-tiered understanding of wholes in the strong sense that some kinds of whole display a three-tiered structure - involving parts, lower-grade whole, and higher-grade whole. Whereas the first interpretation takes him to be suggesting rather an understanding of wholes in terms of two distinct kinds of a two-tiered structure – one kind involving parts and a lower-grade whole, and the other kind involving parts and a higher-grade whole.

Aristotle's distinction between substances and heaps does not share this ambiguity, for it can be understood only along the lines of the first reading. Recall the point that it categorizes any given whole as a substance or a heap in accordance with whether the parts of the whole do or do not (respectively) bear relations of ontological dependence to one another. Thus in any particular case of a whole, since the parts cannot both bear and not bear such relations to one another, the parts cannot compose *both* a substance *and* a heap. Figure 3 is meant to illustrate this view. Again, the squares represent parts, and the single arrows represent the relation of parthood. The divided rectangle represents a heap, Aristotle's version of a lower-grade whole. The undivided rectangle represents a substance, his version of a higher-grade whole. The dotted and continuous double arrows represent relations between the parts which do not, or do, involve ontological dependence (respectively).

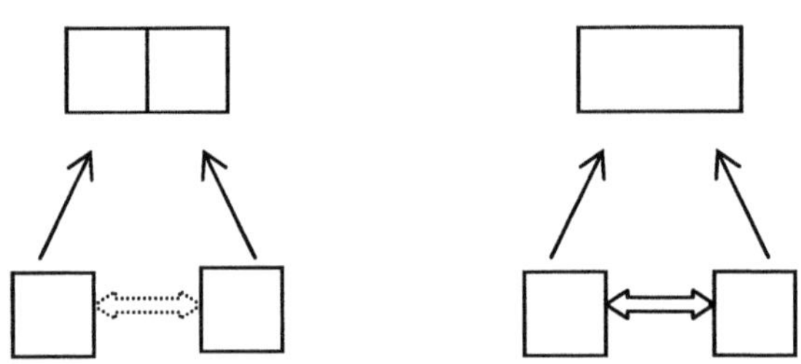

Figure 3

We may therefore view Aristotle's account as interpreting Plato's (ambiguous) proposal by neglecting the possibility of three-tiered wholes in the strong sense, and by explicating the distinction between two fundamentally different kinds of two-tiered wholes. The resulting picture is one in which Plato's most serious problem has largely disappeared. For now we can identify the whole, and apparently adequately characterize it, by identifying its parts and characterizing their properties and the relations between them. We may indeed admit that the whole is different from its parts, but we may take this difference to be the arguably non-mysterious one that while it is one, they are many, and while it has each of them as one of its parts, they do not (presumably S is not *a part of* the two parts S and O; rather, it is simply *one of* them). An organism, then, is adequately identified and characterised by saying that it is the upper level in a two-tiered structure of the 'substance' sort, a structure involving many ontologically interdependent organs and one whole which has those organs as parts. If, as suggested by the alternative interpretation of the *Theaetetus* passage, an organism were assumed to be the upper level in a *three*-tiered structure - involving many organs, a whole identical with those organs, and a whole different from them – then we could not hope to identify and characterize it adequately in this manner. For by specifying the organs as its parts, and the relations between them, we would not have distinguished between the organism and the *middle* level whole which is involved in the same structure (the wholes at both levels have the same parts, related to each other in the same way), and thus we would not have determined the difference between the whole and its parts.

Aristotle's theory of substance allows us, therefore, to formulate a response to Plato's challenge, one which implies a particular approach to the task of providing an account of wholes. According to this approach, which we may call the 'Aristotelian' approach, a whole is completely accounted for by

(1) identifying its parts,
(2) specifying the properties of these parts, and
(3) specifying the relations between them.

Often the main emphasis will be on the latter aspect of this undertaking, namely, the specification of the inter-part relations, either because the identity and properties of the parts are thought to be well understood, or because their identity and properties are thought to be in some way determined by the relations which they bear to one another (as in the case of the parts of Aristotelian substances). The approach presupposes that wholes are two-tiered structures, although it accepts that there may be fundamentally different kinds of such structures, distinguished by fundamentally different kinds of parts or fundamentally different kinds of relation between the parts. Furthermore, it implies that the character of the lower level of the two-tiered structure determines the character of the upper level, in that facts about the identity, properties and interrelations of the parts determine the facts about the properties of the whole.

Considering this approach with Plato's *Theaetetus* passage in mind, however, we see how an objection to the Aristotelian approach might be formulated. For it may be claimed that an account which identifies the parts and specifies their properties and the relations between them cannot be held to be a complete account of the whole. Such an account neglects to specify the relation between the *whole* and the parts (as contrasted with the relations between the parts themselves), and this is a crucial flaw of the Aristotelian approach because *different* relations between the whole and the parts are possible. The assumption that different whole-parts relations are possible might be motivated by the conviction that, *pace* Aristotle, some wholes display a three-tiered structure, involving parts, a lower-grade whole, and a higher-grade whole, or indeed, that some wholes display a many-tiered structure (involving parts, a first-level whole identical to the parts, a second-level whole distinct from the parts, a third-level whole distinct both from the parts and from the second-level wholes, etc.). In fact, the assumption that different whole-parts relations are possible does not imply the rejection of the two-tiered view, and it is conceivable that other justifications might be offered for this assumption. As long as it is embraced, one is bound to reject the Aristotelian approach as to what constitutes an account of a whole. An alternative approach might be proposed as follows: The provision of a complete account for a whole involves

(1) identifying its parts,
(2) specifying the properties of these parts,
(3) specifying the relations between them, and
(4) specifying the relations between the whole and the parts.

To the extent that the identity of the parts, their properties, and the relations between them are thought to be well understood, the emphasis in such an approach is likely to be on the relations between the whole and the parts. It will be convenient to designate this as the 'Platonic' approach, with the understanding that this designation does not imply that the approach is the one Plato subscribed to, but rather, less committedly, that it is an approach suggested by reflection on the *Theaetetus* passage

We have, then, two fundamentally different approaches to wholes, to what counts as an adequate account of a whole, the Aristotelian and the Platonic. They reflect fundamentally different understandings of the nature of wholes. The Platonic approach is compatible with the assumption that some wholes are three-tiered structures (or more generally, that some wholes are many-tiered structures[8]), whereas the Aristotelian approach is not. The former considers that wholes may differ from one another as it were along a dimension which the latter does not envision. For different relations between a whole and its parts may be described as different *vertical* relations within the tiered structure associated with the whole. Thus, if a whole is associated with a three-tiered structure, the lower-grade whole and the higher-grade whole stand in different vertical relations to the parts. The Aristotelian approach tends to take the most interesting aspect of a whole to be the relations between its parts, which we may describe as *horizontal* relations. The Platonic approach will point out that this is to neglect the urgent need to explain the vertical relations, and that these may turn out to be more interesting and consequential than the horizontal ones. From the Platonic point of view, the Aristotelian approach involves the assumption that the facts about the parts – about their identity, their properties, the relations between them – determine the properties of a whole in an overly strict or rigid way. The nature of wholes allows for a certain freedom, relative to the parts, to which the Aristotelian approach is blind. As a result of this blindness it is possible that many of the familiar sorts of entities in the world (perhaps organisms, or sentences, or stories, or melodies, or historical events) will be conceived in a distorted, impoverished manner, by someone who adopts this approach.

I have said that the Platonic approach differs from the Aristotelian in the way it pays attention to vertical relations in wholes. This difference can be stated more precisely by contrasting the assumptions implied by these approaches regarding the relation of *composition*. I have casually used the term 'composition' above, but we now need to be clearer about

---

[8] It is sufficient for my present purposes to discuss only two- and three-tiered structures. I shall return to the possibility of structures involving more than three tiers in the third part of the book.

its meaning. Consider a particular token of the name 'Socrates', a whole which has eight letter tokens among its parts (we may take it that it has additional parts – for instance, a token of the syllable 'tes', or the upper half of the 'c' token – but none of these is a letter). We say that the letters *compose* the word, and in doing so imply that the letters *are parts of* the word. However, composition and parthood are distinct relations. The first three letters are parts of the word, but do not compose it. Each of the letters is a part of the word, but none of them compose it. Composition is a relation which all of the eight letters taken together, *collectively*, bear to the word, while parthood is a relation which all the letters taken separately, *distributively*, bear to the word (whereby each of them bears it individually to the word). Later, in the course the present study, I will have more to say in the way of explicating the difference between these relations, but for the present intuitive purposes these straightforward and uncontroversial observations will suffice.

What the Platonic approach suggests, and what it maintains that the Aristotelian approach ignores, is that there might be different *kinds* of composition, that by saying that the eight letters compose the name we do not fully specify the relation between the letters and the name. To take the example of the syllable SO again, this might be a three-tiered structure, in which case the letters S and O compose both a lower-grade and a higher-grade whole, bearing one kind of relation of composition to the lower-grade whole, and another kind to the higher-grade whole. In whatever way you analyse the relation of parthood, the Platonic approach urges, you will have to acknowledge that different conceptions of composition are compatible with that analysis. If 'compose' is to be defined in terms of 'is a part of', then two distinct definitions will have to be provided, corresponding to the two distinct wholes composed by the same parts. Subscribers to the Aristotelian approach may disagree as to the precise characterization of parthood (for instance, they may disagree whether parthood is a transitive relation or not), but they will share the assumption that composition is uniquely analysable in terms of parthood, that 'compose' is uniquely definable in terms of 'is a part of'. The Platonic approach involves the rejection of this assumption.

It is the purpose of the present study to argue for a Platonic approach as offering a new and fruitful way to treat important problems which arise in thinking about wholes and parts, one which departs from fundamental assumptions shared by the main trends in current writing in the field. It seems to me fair to describe these trends, broadly speaking, as

Aristotelian, in the sense explained above. In describing them thus, I do not intend to be understood as making the historical claim that a direct line of evolution can be traced between Aristotle's theory of substance and the various conceptions of wholes which can be found in present day metaphysical literature, a claim which, if sufficiently supported, would perhaps justify speaking of an 'Aristotelian tradition'. This is not a historical study, and historical discussion will be confined to what seems to me to be required for explaining the theory I am concerned with presenting. Still, it may be noted that in recent writing Aristotle is often discussed and widely viewed as an authority on wholes and parts, whereas Plato is not. An anecdotal indication of this is that the most important contemporary book-length treatment of the field, Peter Simons' *Parts* (a book to which the present study is profoundly indebted), has twenty one page-references under 'Aristotle' in the index, while none under 'Plato'. Ironically, though, from the present perspective, a quotation from the passage of the *Theaetetus* discussed above serves as the book's motto. In view of my characterization of the Aristotelian approach as focusing on parts – their identity, their properties, the relations between them – in its account of wholes, and as largely neglecting ways in which the associated facts about the parts fail to determine the nature of a whole, it is perhaps a significant coincidence that Simons' book on the theory of wholes and parts is called 'Parts'. And it may be similarly indicative that the generally accepted term for theories of wholes and parts is 'mereology', which derives from '*meros*', the Greek word for 'part'.

The Aristotelian character of contemporary understanding of wholes and parts can be seen by looking briefly at some of the major preoccupations of discussion in this area in recent years. The need to clarify one's views about wholes and parts arises in connection with some of the most intensely discussed issues in contemporary metaphysics, including identity, vagueness, constitution, emergence, reduction, supervenience, ontological dependence, composition, set-theory, universals, substance, and temporal parts: (a) In treating the identity of concrete entities over time, one faces problems associated with the assumption that entities can preserve their identity in spite of loss or gain of parts. (b) As for vagueness, the suggestion that it is not only a linguistic phenomenon but also an ontological one is usually understood as the claim that concrete entities typically have vague boundaries, which is a way of saying that in such cases some things are indeterminately parts of the entity. (c) The theory of constitution concerns the relation between a concrete entity and the material (the lump of matter) it is made of, for instance a statue and the clay it is made of, and involves the problem that if, as many maintain, the constituting lump and the constituted entity are distinct from one another, it should be possible for the same parts to compose distinct entities at the same time. (d) The issues of emergence, reduction and supervenience principally concern questions about the way the properties of parts and the relations between them relate to the properties of the whole: Whether the latter are emergent with respect to, reducible to, or supervenient upon, the former (respectively). (e) The notion of ontological dependence, according to which the existence of one entity may depend necessarily on the existence of another, is often discussed in connection with the study of ways in which the parts of a whole may be related to one another. (f) Questions about composition (a relation briefly explained above), and particularly about the sort of relations which two or more entities must bear to one another in order for them collectively to compose a whole, have been given much attention in recent literature, and are obviously directly concerned with the theory of wholes and parts. (g) An attempt has recently been made (by David Lewis in *Parts of Classes*) to provide a 'naturalized' account of set-theory by interpreting it in terms of the notions of whole and part. (h) The study of universals has involved considerations regarding relations between wholes and parts, particularly in connection with D.M. Armstrong's theory of structural universals. (i) A similar point applies in connection with theories of substance which attempt to explain the relations between a substance and its features (properties and relations), especially the bundle theory of substance. (j) Finally, the study of the temporal aspect of reality, of temporal characteristics of entities and processes, extensively involves the notion of a temporal part.

A consideration of such metaphysical issues in recent decades has led to the formulation of a variety of theses, often controversial, concerning the nature of wholes and the relations between wholes and parts. And the point I wish to suggest is that these theses have been formulated, elaborated and discussed within the framework of an Aristotelian approach to wholes and parts. I have in mind claims such as the following: (a) That the nature of concrete wholes is correctly described by the assumptions of classical extensional mereology, a theory which will be introduced and discussed below (the claim was upheld by Nelson

Goodman and David Lewis, for example).[9] (b) That wholes have their parts essentially (this is Roderick Chisholm's well known thesis of mereological essentialism).[10] (c) That wholes may change with respect to the parts they have, i.e. may gain or lose parts (a claim usually made by those who reject both (a) and (b)).[11] (d) That the existence of a whole depends on there being relations of a certain kind between the parts (Peter van Inwagen's recent detailed account of such relations has generated much interest in this connection).[12] (e) That the parts of some wholes exhibit relations of ontological dependence to one another (the modern version of this view stems principally from Husserl's account in the third of his *Logical Investigations*, and has been developed in the works of Peter Simons).[13] (f) That distinct wholes may have precisely the same parts at the same time, or in other words, that composition is not unique (a claim put forward by David Wiggins and adopted by many of those concerned with the identity of concrete entities).[14] (g) That a whole is necessarily distinct from its parts (this is widely, though not universally, accepted, even by authors who otherwise have starkly contrasting conceptions of wholes and parts, such as Lewis and van Inwagen; note that it in effect decides between the two options proposed in the passage from Plato's *Theaetetus* discussed above).[15]

It may be remarked, regarding these theses, that none of them mentions the possibility of that there might be different kinds of relation between the parts of a whole, on the one hand, and the whole, on the other hand (and in particular different kinds of composition), and indeed, that none of them expresses any obvious interest in such a possibility. Rather, they may be broadly characterised as being concerned, on the one hand, with the identity of parts and the relations between them (and especially with the way their identity and the relations between them serve as conditions for the existence of the whole – theses (a)-(e)), and on the other hand with

---

9 See Goodman 1951; 1956; Lewis 1986a.

10 See Chisholm 1976; Simons 1987, Ch.7.

11 See Simons 1987, Ch.5; van Inwagen 1990.

12 See van Inwagen 1990; 1993; Sanford 1993; McGrath 1998; Eklund 2002. Eklund claims (p.245) that 'Peter van Inwagen's *Material Beings* is easily one of the most influential books on metaphysics of the 1990s'.

13 Husserl 1901; Simons 1982; 1987 Ch.8.

14 See Wiggins 1968; 1980; Doepke 1982; Johnston 1992; Lowe 1989, 1995.

15 See Lewis 1991, 80-87; van Inwagen 1994.

the nature of a presupposedly single sort of composition (especially as regards the question whether it implies the uniqueness of the composed whole, and whether it implies that a whole is distinct from its parts – (f) and (g) respectively). The former concern obviously falls within the Aristotelian framework, which I described as accounting for wholes in terms identity of parts, their properties, and the relations between them. The latter concern is consistent with the Aristotelian approach in considering different ways to explicate composition formally, while assuming that only one such explication is correct. Thus, even when Simons provides a formal account of composition according to which two or more wholes might be composed of precisely the same parts at the same time, he supposes that the relation between the parts and the first whole is the same as the relation between the parts and the second whole (in other words, that the parts bear the same non-unique sort of composition to each of the wholes).[16]

In arguing for a Platonic approach, therefore, I take myself to be criticizing the predominantly Aristotelian framework within which contemporary discussion of wholes and parts is conducted. Indeed, my argument will be constructed as a presentation and critique of the Aristotelian framework as embodied in contemporary views, followed by the presentation of a Platonic alternative and the demonstration of insights and theoretical advantages which it offers. More specifically, I shall argue that a wide variety of views share a certain formally statable conception of wholes as 'sums' (which I take to be expressive of an Aristotelian approach), point out some of the more important weaknesses of this conception, and present an alternative conception in the shape of a formal theory which allows for wholes to be either 'sums' or 'Unities' (where the two kinds of whole involve two distinct kinds of composition, as required by a Platonic approach). The sense in which I use the terms 'sum' and 'Unity' will of course be explained as we proceed. My direct concern will be to show that the Aristotelian framework is ill-suited to resolve perplexities and puzzles which arise when we attempt to understand the concrete aspect of reality, the realm of concrete entities, in terms of the notions of whole and part (where my focus will be on problems associated with the issues of identity, composition, constitution, and vagueness), and

[16] See Simons 1987, 26-37, 112-117; I argue in the last section of Chapter 5 that this point holds in spite of the distinctions between kinds of sum which Simons draws on pages 177-187.

that a Platonic framework is better suited to do so. Indirectly, however, my aim is not merely to provide a theoretical response to this particular family of perplexities and puzzles, but rather to outline an approach which seems to me to allow us to articulate intuitions about wholes and parts, intuitions about the richness and elusive character of wholes in many different domains, which the Aristotelian approach tends to stifle.

As a way of suggesting such intuitions, and thus hinting at the more general motivations behind the present undertaking, I now turn to the presentation of certain perceptual phenomena which I hope can help the reader to entertain the idea of a three-tiered whole in concrete terms. Recall that the Aristotelian approach to wholes and parts must reject such an idea, and that its acceptance calls for a Platonic approach. It will be a nice confirmation of this idea, if a theory of wholes and parts compatible with it is shown to offer solutions to the metaphysical puzzles hinted at above. This book, by and large, is devoted to showing this. Before embarking on this project, however, it will be helpful to enhance the *prima facie* appeal of the idea of a three-tiered whole by means of the following illustration.

III - Illustration of the idea of a three-tiered whole

Consider a straight row of twelve golf-balls laid out before us, from left to right. The distances between the balls will influence the way we will tend to describe them. Let us first arrange them so that adjacent balls are four inches apart from one another. Next, let us change the arrangement, drawing the first and second ball three inches closer to one another (so that now they will be only one inch apart), and similarly with regard to the third and fourth, the fifth and sixth, and so on (see Figure 4). It is natural to describe the contrast between these two cases by saying that whereas in the second we have six pairs of balls, in the first we simply have twelve balls, or (to emphasize the difference) twelve single balls.

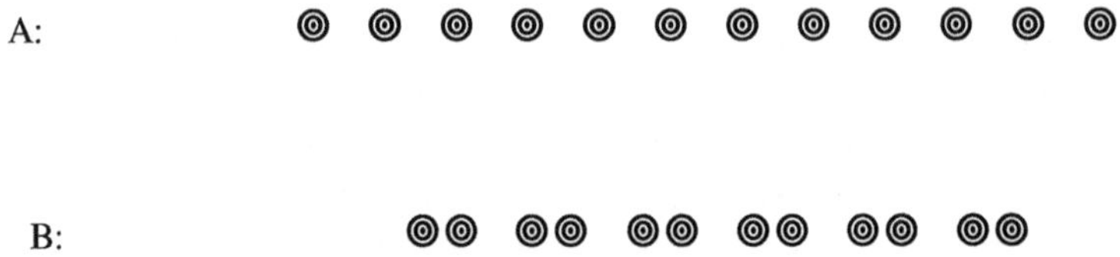

Figure 4

By means of analogous alternative readjustments of the distances between the balls, we will obtain the following further possible arrangements (Figure 5):

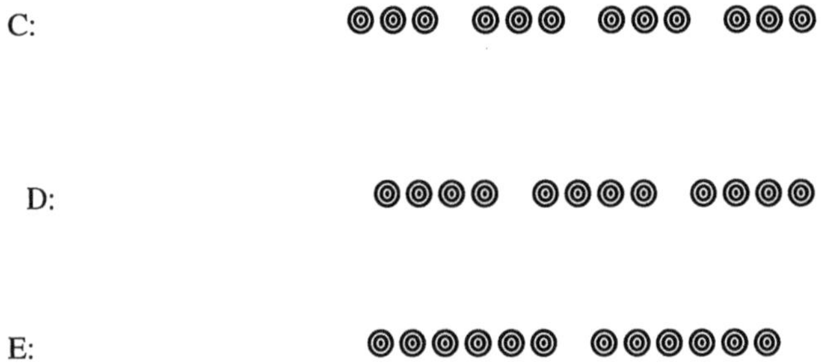

Figure 5

The last three arrangements can be distinguished as cases in which we have four triplets, three quadruplets, and two sextuplets of balls, respectively. By changing the distances between the balls in these ways, we have induced a change both in the way we tend to *describe* and in the way we tend to *perceive* what is before us. In the case of each of the arrangements B to E we perceive the balls as forming larger units, or groups, or indeed wholes (in a sufficiently non-technical sense of the term), and we do so differently in each of the cases, obviously in accordance with the different cues given to us by the relative distances between the balls. The sequence of arrangements from A to E may therefore be viewed as one in which the balls are seen as forming progressively larger perceptual units. In A the unit is a single ball, in B a group of two balls, in C a group of three, in D a group of four, and in E a group of six balls. Such grouping into perceptual units is often prompted by the arrangement of the balls, by such things as the relative distance between them. But as a matter of experience the grouping may vary without the arrangement changing in any way. Thus it seems that most people will be able, when asked, and given sufficient concentration, to look at the balls arranged as in A, and come to perceive them as grouped either in pairs (groups of two), in triplets (groups of three), in quadruplets (groups of four), or in sextuplets (groups of six). We seem to be able to shift at will our perceptual perspective in this sense.

The progression of arrangements from A to E suggests one further step, completing a natural sequence. In changing the arrangements we left progressively fewer four-inch gaps between the balls – eleven such gaps in A, five in B, three in C, two in D, and finally one in E. Let us now 'close' this final gap, giving us a contracted, uniform row of balls, F, and let us consider this last arrangement in the context of the previous ones by surveying the whole sequence at a glance (Figure 6). Since we describe A as twelve single balls, B as six groups of two balls, C as four groups of three balls, D as three groups of four balls, and E as two groups of six balls, it is only natural to describe E as one group of twelve balls.

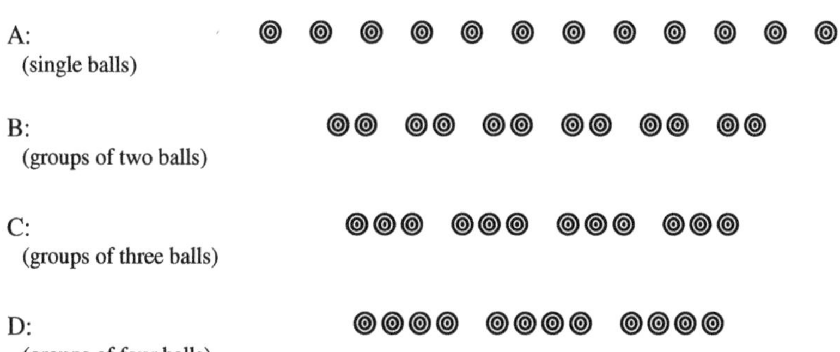

E: ◎◎◎◎◎◎ ◎◎◎◎◎◎
(groups of six balls)

F: ◎◎◎◎◎◎◎◎◎◎◎◎
(a group of twelve balls)

Figure 6

The point of presenting this sequence can now be revealed as that of focusing on the contrast between A and F. I have said that each of the arrangements A to E tends to be *perceived* differently, as regards grouping. When we consider F, we are easily convinced that it represents a different grouping, a way of perceiving the balls which is different from the one associated with either B, C, D or E. Comparison between F and A is more tricky, however. It is easy enough to conflate them, to think that as regards the grouping aspect of perception they must be the same. I suggest that this is not so. When we move up the sequence towards A, we move towards progressively *less* cohesive ways of perceiving the twelve balls, whereas when we move down the sequence towards F, we move towards progressively *more* cohesive ways of perceiving them. When we perceive the many balls, the content of our perception can be less or more cohesive, less or more thoroughly grouped together, and A and F represent the minimal and maximal ends of the range of this variation. A is perceived in a completely 'ungrouped-together' way, and F is perceived in a completely 'grouped-together' way. To this difference corresponds our describing them as twelve (single) balls, and one group of twelve balls, respectively.

It should be emphasized that the difference between the two ways of perceiving the balls associated with A and F does not depend on the change in the distances between the balls. Had we started by arranging the balls so that the distance between each and the next were twelve inches instead of four, progressing by steps (involving a shortening of distance between balls to four inches) to arrangements perceived in increasingly cohesive ways, we would have constructed a sequence similar to the one we have been discussing. In that case, however, we would find A (with balls separated by four inches) at the *bottom* end of the sequence. Thus, it would be perceived in a *maximally* cohesive way, whereas in the former case it was perceived in a *minimally* cohesive way.

I am claiming, therefore, that when we are presented with a row of twelve evenly spaced balls, we can perceive them in two contrasting ways which might be characterised as minimally cohesive and maximally cohesive, and that we are to a large extent at liberty to switch from one way of perceiving them to the other. It will be convenient to use shorter expressions: I shall describe a minimally cohesive perception as 'fragmented', and a maximally cohesive one as, simply, 'cohesive'. In the former case we perceive them as *many* (i.e. many balls), while in the latter as *one* (i.e. one group of balls). Our ability to perceive them as one seems to be associated with our ability to count groups. Had we arranged balls in rows on three tables (say, twelve balls on each table), for example, we could have counted three groups of balls. Doing so seems to imply thinking of the twelve balls on each table as one.

Note that an important feature of the contrasting ways of perceiving the balls is their incompatibility with one another. We can perceive the balls as four triplets, and we can perceive them as three quadruplets, but we cannot perceive them in both ways *at the same time*. If we start with perceiving them as triplets, our perceptual experience must undergo a *switch* in order for us to come to perceive them as quadruplets, much in the way it must when we perceive the figure which Wittgenstein made famous, the duck-rabbit, alternatively as duck and as rabbit. And this, I submit, is also the case when we alternate between perceiving the balls as twelve single balls and as one group of twelve balls. For entities to be many is *perceptually incompatible* with their being one. I mentioned earlier that authors widely, though not universally, agree that the idea of one-many identity is incoherent (and thus that 'SO' and 'S and O' cannot have an identical reference). My claim here is different: I am saying that even if it is possible and indeed conceivable that one entity should be identical to many entities, being one is *perceptually* incompatible with being many. In other words, whether or not *being* one is compatible with being many, *appearing* (perceptually) to be one is incompatible with appearing to be many. This is a remarkable characteristic of the features of being many and being one. Often clearly distinguishable features are compatible with one another. For example, being orange is perceptually compatible with being round. It is the perceptual incompatibility between being many and being one which seems to make it so natural to describe cases displaying both features in terms of a two-tiered structure, where the tier below corresponds to being many and the tier above corresponds to being one. The notion of tiers, of distinct levels, suggests the idea that what we perceive cohesively lies above, and conceals, as it were, what we perceive fragmentedly, making it impossible to perceive in both ways simultaneously.

The distinction between fragmented and cohesive perception seems to be a general one, not confined to simple and symmetrical cases such as those we have been considering. We might have taken a smaller or larger number of balls, we might have arranged them in various two or three dimensional patterns, we might clustered them together or spread them out, and indeed, we might have replaced them with objects of different kinds.[17] Thus we might have arranged various vegetables in a bowl, and set about to perceive them now in a fragmented way, now in a cohesive way. It seems that whenever we arrange many common objects (as long as we can see all of them at the same time, and distinguish them by sight from one another), we can perceive them either fragmentedly, as many, or cohesively, as one.

It might be supposed that this structure tells us all there is to know about the way in which the appearance of being one can feature in cases where we perceive many things. I wish to argue that this impression is mistaken, and that in some cases a third tier can be clearly indicated, distinct from the second in spite of sharing the feature of appearing to be one. The conclusion will be that a three-tiered structure is better suited than a two-tiered one to describe what we perceive in such cases.

Consider the painting *The Vegetable Gardener* by Giuseppe Arcimboldo (c.1527-1593), which depicts a bowl overflowing with a variety of vegetables (Figure 7). If the observations made above are correct, it should be possible to perceive the image either fragmentedly or cohesively. In the former case we perceive it (i.e. what is depicted in the painting) as many objects, which we describe as a bowl and varied vegetables. In the latter case we perceive and describe it as one group of objects, a group which includes a bowl and varied vegetables (Aristotle might have preferred to describe this sort of group as a 'heap'). These two ways of perceiving the image would be equally available if we tilted the image in different ways. So far, we have no reason to think that

---

[17] I use the term 'object' in the sense of 'everyday object', things such as golf-balls and vegetables. As a term for designating anything whatsoever I use 'entity'. My discussion throughout this book will be focused, by and large, on a limited class of entities, but a class which still includes many entities which are not everyday objects (e.g. an elementary particle; a group of golf-balls).

accounting for what we perceive in terms of a two-tiered structure is in any way inadequate.

Figure 7

However, if we turn the painting on its head, something remarkable happens (see Figure 8). Suddenly a switch occurs in our perception, and a grotesque face appears. We may, with some effort, switch back to perceiving the image as a group of objects (bowl and vegetables), but then the face fades away. The appearance of a group is incompatible with the appearance of a face.

Figure 8

Clearly, the changes in the way we perceive the painting do not correspond to changes in the properties of the actual painting, and particularly changes in the relations between the parts of the painting corresponding to the various objects. And significantly, we report the perceptual change as a change in the way we perceive what is before us, but *not* as a change in the way we perceive the *relations between the vegetables*. Perceiving something as a face is incompatible with perceiving it as a group of vegetables, and in particular with perceiving it as a group of vegetables related in a new way.

But if perceptual incompatibility invites description in terms of tiers, then in the case of Arcimboldo's painting we are led to distinguish *three* tiers. For the case involves first of all the perceptual incompatibility between the many objects and the group of objects (which is the incompatibility between fragmented and cohesive perception), and second, the perceptual incompatibility between the group of objects and the face. We therefore have here many objects which can be perceived in any one of three distinct ways: (1) as many objects (the bowl and vegetables), (2) as one group of objects (which includes the bowl and the vegetables), and (3) as one object (the face).

Note that while (1) involves perceiving the many objects *as many*, (2) and (3) both involve perceiving them *as one*, so that two ways of perceiving them as one must be distinguished from one another. I have called the way of perceiving associated with (2) 'cohesive'. Let us say that (3) is associated with a 'unitive' way of perceiving. Thus, the perception of such cases is characterised in terms of three distinct tiers, fragmented, cohesive, and unitive. How should we explicate the perceptual difference between the two ways of perceiving the objects as one, that is, the difference between cohesive and unitive perception? In other words, what is the difference between perceiving them as one group of objects, and as one object? It seems that at least a tentative answer can be given as follows: In the case of perceiving many objects (e.g. vegetables and a bowl) cohesively, the notion that those objects are included in what we perceive is an essential feature of the way it appears. We perceive the objects *as* one thing which includes those objects. This is reflected in our using the term 'group' to describe this case, for a group is always thought as a group *of* several or many things. In the contrasting case of perceiving many objects unitively, that notion ceases to feature in the perception, and consequently we have the sense that the objects have 'disappeared': Seeing the face, we no longer seem to see a bowl and vegetables. Appropriately, we do not use the term 'group' in this case. A group is thought of as a group *of* things. A face, of course, is not thought of as a face 'of things' in any similar way.

Once it is acknowledged that accounting for our perception of this painting needs to be done in terms of a three-tiered structure which displays the relations between fragmented, cohesive, and unitive ways of perceiving, it may be realized that such a structure is required not only in connection with eccentric or contrived cases of this sort. For I suggest that the peculiarities of Arcimboldo's pun do not introduce an incompatibility (between the cohesive and unitive levels) which is absent in more naturalistically rendered portraits. The virtue of the painting's peculiarities is that they attract our attention to an incompatibility which is present though less striking in many more mundane cases. The incompatibility between cohesive and unitive perception forces itself upon us in the case of *The Vegetable Gardener*, because our concept of a man (both according to common sense and according to science) makes it *obvious* that what appears as a group of vegetables and bowl cannot appear as a face; for to appear as a face something must appear as not being made up of vegetables and bowl. In the case of a more typical portrait, what corresponds to the group of vegetables and bowl is perhaps a group of facial parts (mouth, nose, eyes, cheeks, brow, etc.). And neither common sense nor science provide us with clear reasons for claiming that what appears as a group of facial parts cannot appear as a face. Our tendency would be presumably to suppose that the appearance as group and the appearance as face, if at all distinguishable, are related to one another in the way appearance as orange and appearance as round are – that is, we would suppose they are compatible features. Reflection on our perception of *The Vegetable Gardener* gives us reason to conclude that this supposition would be mistaken. It helps us to realize that what we experience as a perceptual switch, in making the transition from perception as face to perception as group of vegetables and bowl, is an operation which has a direct analogue in the case of a typical portrait. And the analogous operation in the latter case is equally experienced as a perceptual switch, indicating perceptual incompatibility.

My discussion of the sense in which perceptual experience seems to require an account in terms of a three-tiered structure has so far been confined to cases in which we perceive paintings. A similar point can be made in connection with many other cases in which we perceive visual representations. Consider, for example, a computer printout of letters arranged in a way which we recognize to represent a bust of Napoleon. Here too it seems possible to switch between perceiving the letters as many letters, as a group of letters, and as a bust. Similarly, in the case of a low-quality newspaper photo-print, we may switch between perceiving many dots, a group of dots, and a figure. The need for a three-tiered account, however, seems to arise equally in connection with the perception of entities more generally, and not only entities in representation. In order to be convinced of this, it will be useful to conduct a minor thought-experiment.

Let us suppose that the laws of nature were such as to make it possible for a strange species of organism to exist, a species much like humans in terms of their behaviour and general appearance, but radically different from humans as regards their physiology, in that they are made up of vegetables, and let us suppose that Arcimboldo's painting depicts a certain member of that species. If we met him and looked in his direction, it seems in view of the foregoing discussion that we might experience any one of three distinct kinds of perception – fragmented, cohesive or unitive. Perception provides us with different appearances in each of these cases: What we perceive in each case appears to be different from what we perceive in any of the other two cases.

It seems therefore that the distinction between fragmented, cohesive and unitive modes of perception is applicable to non-representational cases, and we can inquire about the way we perceive particular examples or kinds of examples in the light of this distinction. Is our typical perception of a jointed cactus better described as cohesive or unitive? Are we not able (when we persistently apply ourselves to do so) to switch between three modes of perception in this case, paralleling our ability in connection with the representational examples just mentioned? And how about a case in which we perceive the legs, arms, trunk and head of a person (belonging to the species *homo sapiens*) standing before us? Do similar distinctions not apply here as well? At least it can be said that such questions make sense. And by asking them, we are inquiring about the world of perception in terms of three-tiered structures.

Indeed, such reflections quickly lead us beyond perception into the realms of imagination. When we considered the printout of Napoleon's bust, we did not actually need to perceive it in order to grasp the distinctions. It was enough to imagine the case. But imagining this case seems to be analogous to imagining others which are not as easily, or not at all, accessible in principle to perception. I have in mind cases such as imagining (i.e. imagining in visual terms) a person as made up of many bodily organs (heart, liver, lungs, brain, bones, etc.); or even imagining a person as made up of innumerable individual cells, or of elementary particles. The question whether we are in such a case imagining the many entities (organs, or cells, or particles) as many, as one group, or as one entity (i.e. an entity which is not a group), is analogous to the question posed in connection with Arcimboldo's painting, and thus seems to involve at least entertaining the possibility that what we imagine should be described in terms of a three-tiered structure. We may not know, as yet, how to explain the differences between the tiers in more precise terms, but presenting them to ourselves for consideration seems to make intuitive sense.

It is now time to return to the notions of whole and part, which are my principal concern in this study. The relevance of the foregoing discussion to the way we think about these notions can be explained as follows. We have been considering different modes of perception, and corresponding ways in which things may appear to us when we perceive them, and we have seen that parallel distinctions might be drawn regarding modes of imagination and the appearances associated with them. Now clearly, the notions of whole and part apply widely to appearances, to the world as it appears. How would they apply in the circumstances we are considering, say, in connection with our hypothetical gardener? In the fragmented case, what we perceive appears to be many entities which we may characterize as parts. In the cohesive case, it appears to be one entity which we may characterize as a whole. And in the unitive case, what we perceive appears to be one entity again, which we may characterize as a second whole, distinct from the first (we may distinguish the two wholes as lower-grade and higher-grade, respectively). Indeed, it may well be that the grounds for applying the term 'part' to characterize what we perceive in the fragmented case is precisely our realizing that by switching to a different mode of perception, what we now perceive as many entities *will* be perceived as one object. Similarly, the grounds for applying the term 'whole' to characterize what we perceive in the cohesive and unitive cases may be the corresponding realization that by switching to the fragmented mode, what we now perceive as one entity *will* be perceived as many entities. For it seems that the terms 'part' and 'whole', as I shall argue in Chapter 4, are *correlative* terms: Assuming that x is a part seems to imply the assumption that some y is a whole, to which x belongs as a part; and assuming that y is a whole seems to imply the assumption that some x belongs to y as one of its parts.

It follows that the hypothetical gardener at least *appears*, as far as perception is concerned, to be associated with a three-tiered structure, involving parts, lower-grade whole and higher-grade whole. And importantly, the difference between the lower-grade whole and the higher-grade one does not appear to be associated with relations between the parts of these wholes. Take parts, and fix the relations between them, and yet two distinct wholes appear, which we take to be equally composed by them.

If the distinction between fragmented, cohesive and unitive modes of perception is valid in connection with non-hypothetical entities as well, such as the jointed cactus, or the man (considered as composed of arms, legs, trunk and head), then we should say that these entities also appear to be associated with a similar three-tiered structure. And finally, if an analogous distinction can be drawn between three modes of *imagining* (as I have suggested), it would seem to follow that various entities which we imagine (such as a man considered as composed of cells) also appear to be associated with a three-tiered structure of parts, a lower-grade whole, and a higher-grade whole.

This conclusion indicates that the Platonic approach better reflects the way the world appears (to perception and imagination) than the Aristotelian approach. The hypothetical gardener, on the one hand, and the group of vegetables and bowl, on the other hand, appear as bearing respectively *different* relations to the bowl and vegetables. As noted, when we perceive unitively, the bowl and vegetables seem to disappear, to be *replaced* by the gardener, which they do not when we perceive cohesively. The group appears as involving the bowl and vegetables directly (again, we describe it as the group *of those objects*), in a way that the man does not. All of this seems to indicate that we have here the appearance of two wholes which bear distinct relations to the same parts. This possibility is affirmed by the Platonic approach, but denied by the Aristotelian approach. An Aristotelian would have to judge appearances to be incoherent in this respect.

How would she interpret the difference between cohesive and unitive perception, then? She might propose something along the following lines: When we perceive unitively, the parts appear as bearing relations of ontological dependence to one another, whereas when we perceive cohesively they do not. When we switch from cohesive to unitive perception, at least one of the two appearances is necessarily *mistaken* or incomplete (for the parts either are or are not ontologically related to one another, necessarily). This seems to be an unsatisfactory solution, however. First of all, it seems not to account for the way the bowl and vegetables seem to disappear or to be replaced, when we switch from cohesive to unitive perception. In what is perhaps the clearest example of ontological dependence, the relation between colour and the shape in the case of a colour-patch (colour cannot exist without shape, nor shape without colour), it cannot be said that when we perceive the patch, the colour and the shape have disappeared or been replaced. Thus, the mere fact that things are perceived as ontologically dependent does not explain their ceasing to appear distinctly in the perception. Second, the Aristotelian response seems to imply that, if we accept the authority of physics regarding the structure of matter, perceiving (or imagining) a man unitively must be mistaken, for physics gives us every reason to believe that the elementary particles composing a man do *not* depend ontologically on one another (any of the particles would continue to exist happily if it were extracted and removed to the other side of the universe). Thus unless the Aristotelian can come up with a different way of interpreting the difference between cohesive and unitive perception, her conception of wholes and parts impels her towards a form of eliminativism which she may well wish to avoid.

I have attempted in this section to provide an intuitive illustration of the idea that wholes may be associated with a three-tiered structure, in the way supposed by a Platonic approach. My intent was to enable the reader to make preliminary sense of this idea in concrete perceptual and imaginative terms, as well as to suggest that this idea helps us both to notice and to interpret certain closely observed aspects of the way in which we perceive or imagine wholes. What I have said here, if convincing, will serve to add motivation to the project to which this book is devoted. Being in possession of a theory which allows us to articulate a certain richness in the way we perceive or imagine the world, or in the way we *may* perceive or imagine the world, is certainly a desirable prospect. However, I shall not be dealing with perception other than incidentally, within the limits of the present study. And my argument will not depend in any way on the claims we have been discussing regarding perception and imagination.

IV - The Theoretical Context

I mentioned earlier in passing that the term which has come to be accepted for designating the theory of wholes and parts is 'mereology' (derived from the Greek for 'theory of parts'). In its widest sense, the term is meant to cover any general assumptions or claims intended to explicate what is involved in or entailed by an entity's being either a part or a whole or both.[18] Somewhat more narrowly it may be understood as designating any formal theory which attempts to articulate the notions of whole and part systematically. However, the term is often understood still more narrowly, in its original sense as the name which Stanisław Leśniewski gave to a theory he proposed in 'Foundations of a General Theory of Sets', a paper published in Polish in 1916.[19] His work was largely neglected until a version of the same theory was presented in 'The Calculus of Individuals and its Uses' by Henry Leonard and Nelson Goodman in 1940.[20] Since then it has gradually become standard to put forward general claims about wholes and parts against the background of Leśniewski's theory, which I will distinguish in what follows as *classical* mereology.[21]

The theory of wholes and parts is formulated in classical mereology by analogy with set-theory, which by Leśniewski's time was occupying much of the attention of mathematicians as well as logicians. Indeed, the notion of a whole which classical mereology expresses evolved in the process of reflection on the nature of sets and on problems encountered in developing set-theory. These problems became critical after Russell discovered the paradox named after him, and Leśniewski's purpose in developing classical mereology was largely to present an alternative to set-theory which would not suffer from the same weaknesses. The character of classical mereology is illuminated by a consideration of its set-theoretical roots.

Georg Cantor's set-theory appeared in its earliest form in 1874, and was gradually revised and developed in a series of publication culminating in his *Contributions to the Founding of the Theory of Transfinite Numbers* in 1895.[22] In this latter work he offered the following famous definition:

---

[18] Thus Desmond Paul Henry's book on medieval views about wholes and parts is called '*Medieval Mereology*' (Henry 1991).

[19] Leśniewski 1916.

[20] Leonard and Goodman 1940.

[21] For a much more detailed presentation of mereology than the one I offer here, see Simons 1987, especially Chs. 1 and 2. For alternative overall characterisations, see Lewis 1991, 72-81, and van Inwagen 1990, 28-32. For more on philosophy in Austria, Germany and Poland in the late 19th and early 20th century as background for mereology, see Smith 1982, 15-65; 1994, especially Chs. 6 and 7.

[22] Cantor 1895; for an extensive study of Cantor and the development of set-theory, see Dauben 1979. See also Tiles 1989 and Pollard 1990. For a much briefer, and less philosophically minded historical survey, see Burton 1985. A very clear and helpful introduction to set-theory can be found in Machover 1996.

By a set we are to understand any collection into a whole *M* of definite and distinguishable objects of our intuition or our thought. These objects are called the elements of *M*.[23]

It is clear from this definition that for Cantor the notions of a set (in German: 'Menge') and a whole ('Ganze') are not mutually exclusive, and that in fact he seems to suppose that a set is some sort of a whole, and that the elements are parts of this whole.

Cantor originally introduced the theory to facilitate reasoning in precise terms about infinite sets of points on the continuous line, which he required for treating a problem he had been working on in the early 1870s, that of the representation of functions by trigonometric series. In devising a theory of sets he took himself to be exploring a notion which is familiar to common sense, and which seems clear enough as long as we speak of everyday, finite sets of objects, and yet gives rise to striking paradoxes when we attempt to speak of infinite sets. In the case of an infinite set of entities *S*, there may be a one to one correspondence between the members of *S* and the members of a proper subset of *S* (i.e. of a set which includes some but not all of the members of *S*). For example, there is a one to one correspondence between the set of natural numbers 1, 2, 3, 4, ... and the set of *even* natural numbers 2, 4, ... Thus paradoxically, there seem to be as many members in the subset as in the set, in spite of the fact that the latter includes all the members of the former as well as members which the former does not include. This feature of infinite sets had already been noticed by Bernard Bolzano, who discussed them (and indeed introduced the term 'Menge' into the mathematical lexicon) in *Paradoxes of the Infinite*, a book published posthumously in 1851.[24] Cantor could not be content with merely observing this paradoxality, for it threatened the coherence of his account of functions. He thus undertook to develop a systematic theory of sets, and one requirement of such a theory would seem to be a clarity about what exactly one takes a set to be. Hence the definition quoted above.

There were good reasons for Cantor to appeal to the notion of a whole in specifying what he means by 'set'. For it seems that the common-sense notion of a set is ambiguous in a way that the common-sense notion of a whole is not, so that defining 'set' in terms of 'whole' may give the

---

[23] See Burton 1985, 588.

[24] Bolzano 1851.

former a less ambiguous and thus more theoretically usable (technical) sense. The word 'set' as used in non-mathematical contexts (e.g. in 'there is a set of plates in the cupboard') belongs to a family of words which grammarians describe as *collective nouns*, a family which includes such words as 'group', 'collection', 'class', 'herd' and 'flock'. It is characteristic of collective nouns that their meaning seems to combine the features of being many and being one in a peculiar way. The word 'whole' is not a collective noun, and although its meaning too seems to combine those features, as I noted in the first section of this introduction, the way in which they are involved in its meaning seems to be different.

For although a whole's having many parts is analogous to a set's having many members,[25] it seems that we tend (in common-sense) to treat a whole as one thing, in spite of the fact that it has many parts, in a way that we do not tend to treat a set. If there is a pepper-grinder composed of twelve parts in the cupboard (and nothing else), my tendency would be to say that there is *one* thing in the cupboard. If there is a set of twelve plates in the cupboard (and nothing else), my tendency would be to say that there are *twelve* things in the cupboard. Our treatment of sets is thus ambiguous: In a context such as the one just described we tend to treat a set as many things, whereas in other contexts we would tend to treat it as one thing (for example, when we count sets of plates). Our treatment of wholes seems not to be ambiguous in this particular way.

In Cantor's theory, it is crucial to uphold a distinction between the set $A = \{a, b, c\}$ and the set $B = \{a, \{b, c\}\}$. A is a set which has *three* members, a, b and c, whereas B is a set which has *two* members, a and the two-membered set $\{b, c\}$. By defining a set as a kind of whole, Cantor indicates that the set $\{b, c\}$ should be treated as *one* member, and not as two members, of the set B.

We can see, therefore, that sets in Cantor's theory are understood as wholes of a sort. However, they are wholes of a very unusual sort. To see why, let us consider fundamental features of sets according to the theory.

(S1) The Principle of Extensionality: Sets comply with the Principle of Extensionality, according to which if each of the members of set A is a member of set B, and vice versa, then A is identical to B; and otherwise A is distinct from B. Sets, therefore, are identified by their members. Wholes, by contrast, do not always seem to comply with such a principle.

---

[25] In mathematical contexts, 'member' is used as equivalent to 'element'; in non-mathematical contexts 'member' is the more typical term.

If the elementary particles which compose a human being were rearranged so as to compose a stone, it would usually be assumed that the stone is *not* identical to the former human being. Furthermore, as I noted in section II above, many contemporary theorists agree that in the case of entities such as a clay statue, the lump of clay and the statue are distinct from one another, and yet the same elementary particles compose them both at the same time. If the lump and the statue are both wholes, this conflicts with extensionality.

In view of the Principle of Extensionality, the theory assumes that a particular set can be uniquely defined by specifying its members – either by listing them, or (more typically) by specifying a property *P* such that all and only the entities which have this property are members of the set.

(S2) The Comprehension Principle: The theory assumes what is known as the Comprehension Principle, according to which *whenever* we specify entities (either by listing them or by specifying a property which they, and only they, have), there is a set which has all of those entities as its only members. Thus, there is a set which has my telephone, the moon and the Empire State Building as its only members. We do not usually acknowledge the existence of wholes defined in such an arbitrary way.

(S3) Non-transitivity of membership: Suppose we defined a set *A* by specifying a property *P*. If one of the members of the *A* is itself a set *B*, then it is by no means guaranteed that the members of *B* have the property *P*. Therefore, the members of *B* need not be members of *A*. Thus, if *A* is the set of religious sects, it has *B*, the set of Manichaeans as one of its members, but none of the individual Manichaeans is a member of *A*. It follows that the relation of membership is non-transitive: If *a* is a member of *B*, and *B* is a member of *A*, then it may but need not be the case that *a* is a member of *A*. Now, the relation of parthood, as it is understood in connection with typical wholes, is usually considered to be transitive. According to this view, if a molecule is a part of the heart, and the heart is a part of the body, then necessarily the molecule is a part of the body. If this view is accepted, therefore, the relation between a whole and each of its members is different from the relation between a set and each of its members.

(S4) Two ways of belonging to a set: Sets in general have *subsets*. *B* is a subset of *A* if each member of *B* is also a member of *A*. If *B* is a subset of, but not identical to, *A*, then *B* is a *proper-subset* of *A*. Sets are said to *include* their subsets. Thus, the set $\{a, b\}$ is a subset of, and is included in, the set $\{a, b, c\}$. If *A* and *B* are sets, then the set of all entities which are members both of *A* and of *B* is called the *intersection* of *A* and *B*. It is important to note that membership and being included in are distinct relations. {*a*, *b*} is a subset of {*a*, *b*, *c*}, but it is not a member of it. Thus, an entity can belong to a set in two distinct ways - either as a member, or as a subset. In the case of wholes, by contrast, it seems that there is only one way in which entities can belong to them – by being parts.

(S5) The null set: The theory assumes that there exists a set with no members at all, called the 'null set' or the 'empty set'. Inconsistency would threaten otherwise. For according to the Comprehension Principle, *whenever* we specify a property, we are guaranteed that a set of entities which have this property exists. But if *no* entity has the property we happen to specify (perhaps no entity is *precisely* spherical), and there is no such thing as the null set, then no set exists which corresponds to that property. It follows from the definition of subsets that *every* set has the null set as one of its subsets. Therefore, *any* two sets, *A* and *B*, have an intersection – they have at least the null set in common. Wholes seem to be different in this respect. We do not think that any two wholes must have an entity which belongs to both of them as a part.

(S6) Distinct singleton: For a reason similar to that which led to the recognition of a null set, it must be assumed that there can be sets which have only one member. If *a* is an entity, the set *A* = {*a*} is called the 'singleton of *a*'. Generally speaking, the singleton of *a* will be distinct from *a*. Thus, if *a* is not a set, then *A*, the singleton of *a*, must be distinct from *a*, by the Principle of Extensionality. For *A* has *a* as a member, whereas *a* is not a set and so in particular does not have itself as a member. Furthermore, if *a* is not a set, then it follows that there is an infinite hierarchy of singletons built upon it – {*a*}, {{*a*}}, {{{*a*}}}, etc. – all distinct from one another. We do not think of wholes as being associated with similar hierarchies.

Gottlob Frege was particularly clear in realizing that Cantor's notion of a set does not correspond to what we usually think of as wholes. In his paper 'A Critical Elucidation of some Points in E. Schröder's Vorlesungen über die Algebra der Logik', published in 1895,[26] he insisted on a distinction between different notions of a class (the terms 'class' and 'set' are synonymous in this context, and 'class' is the term we find more usually in works dealing with logic). On the one hand there is the notion of a 'class as the extension of a concept'. Cantor's theory should be

[26] Frege 1895.

viewed as an attempt to provide a detailed account of classes of this kind. And logicians interested in the logical formalization of language, as Frege was, should be concerned with such classes. On the other hand there is the notion of a 'collective whole', associated with our description of everyday concrete entities as wholes. Frege suggested that the distinction turns on the different ways in which entities are assumed to belong to the two kinds of class, respectively, a point we have just noted in (S4) above. There are two ways in which an entity can belong to a class as extension, whereas there is only one way in which an entity can belong to a collective whole. Later, in the course of my discussion in Part One of the present study, I shall return to Frege's account in more detail.

At the time, the attention of mathematicians and philosophers concerned with sets or classes was indeed often focused on such classes as extensions. Prior to Leśniewski's work on mereology, nothing like a systematic formal theory treating wholes, parts and the relation of parthood, was available, which would parallel set-theory's treatment of sets, members and membership. Still, there was no lack of interest in what Frege would describe as collective wholes, and especially in drawing distinctions between fundamentally different kinds of entities of this sort. As I noted in sections I and II above, the idea that we can distinguish between lower-grade and higher-grade wholes can be found in the works of both Plato and Aristotle. Modern versions of this idea (which may well be described as a perennial philosophical theme) can be found in discussion, around the turn of the century, of the notions of an 'organic whole', of a 'Gestalt', and of a 'pregnant whole' – all of which may be considered as terms for a higher-grade whole.

Organic wholes were discussed by the British Neo-Hegelians and some of their critics, such as G. E. Moore (in *Principia Ethica*).[27] The notion of a Gestalt is associated with theorists influenced by Franz Brentano, and particularly with Christian von Ehrenfels, whose 1890 paper 'On Gestalt Qualities' was particularly responsible for the currency of the term, and with the founders of Gestalt psychology, Max Wertheimer, Wolfgang Köhler, and Kurt Koffka.[28] The notion of a pregnant whole is explored in the third of Edmund Husserl's *Logical*

[27] See Moore 1903a.

[28] See Ehrenfels 1890; Wertheimer 1925; Köhler 1929; Fuller 1990; for a historical survey, translations of works by Ehrenfels, and related essays, see Smith 1988.

Investigations.[29] What the approaches of these varied thinkers have in common is an emphasis on special kinds of relation between the parts of a whole as accounting for the distinction between their respective favorite versions of a higher-grade whole and the contrasting lower-grade whole. Husserl arguably offers the most profound analysis in this connection, in his characterization of the relevant special kind of relation as one of 'foundation' (which, as Peter Simons explains, may be understood as a relation of ontological dependence),[30] and setting out in detail various ways in which this relation can feature in a whole.

Meanwhile, set-theory was compelled to confront the threat posed to it by the startling discovery that its basic assumptions lead to contradictions. Among the most familiar and important of these contradictions was pointed to by Bertrand Russell in the *Principles of Mathematics* (1903),[31] and is known as Russell's Paradox. Cantor's assumption of the Comprehension Principle implies in particular that the class (where 'class' is used for 'set') *A* of all self-membered classes exists (a self-membered class being a class which is a member of itself). The question arises whether *A* itself is one of those classes. Russell showed that if it is assumed that *A* is self-membered, then it follows necessarily that it is *not* self-membered, whereas if it is assumed to be not self-membered, then it follows necessarily that it *is* self-membered. Russell's conclusion was, in the first place, that we cannot conceive of the class *A* as *one* thing, which might then serve as one of the members of some other class (recall that, as I remarked above, it was important for Cantor to emphasize that a set is treated as one thing when it is itself a member of some other set, a point which is suggested by his definition of sets as wholes of a sort). He proposed that we draw a distinction between two notions of a class – a 'class as one', and a 'class as many'. We may legitimately speak of *A* as long as we take it to be a class as many.

The Comprehension Principle, however, is clearly a principle which guarantees the existence of classes in the sense of a class as one. It must be rejected therefore, in its unrestricted form. In the *Principia Mathematica* (1910),[32] together with Alfred N. Whitehead, Russell

---

[29] Husserl 1901; see also Simons 1982; 1987 Ch.8.

[30] See Simons 1987, p.292.

[31] Russell 1903.

[32] Russell and Whitehead 1910.

elaborated a theory which orders all classes in a hierarchy of 'types', and assumes that a class of a certain type can only have classes of lower types (or individuals which are not classes) as members. Therefore, in particular, no class is self-membered (and Russell's Paradox is avoided, as can be readily seen). An alternative approach to restricting the Comprehension Principle was presented by Ernst Zermelo in 1908,[33] according to which set-theory is developed *axiomatically*, that is, as a theory which is built up by deduction from a limited number of unproven axioms. The unrestricted Comprehension Principle is replaced by particular instances of it, plausibly regarded as harmless, which are stated among Zermelo's axioms.

Leśniewski's classical mereology may be understood against this background as a component of the project he undertook to provide a foundation for mathematics without resorting to the problematic set-theory.[34] Following Frege he characterised the sets of set-theory as 'distributive classes', in contrast with 'collective classes' which correspond more closely to the concrete wholes of common sense (the expressions are equivalent to Frege's 'classes as extensions' and 'collective wholes', respectively). He considered that reference to distributive classes can be eliminated, and that the only theory of classes we need is a theory of collective classes – which is precisely what his mereology claims to be. The intuitive point of departure is similar to that of Cantor's set-theory. We suppose there are several or many individuals (the individuals which originally concerned Cantor were points on a continuous line), and take it that there is one thing, some sort of a whole, to which those individuals belong. Cantor's term for the one thing is 'set', and the individuals belong to the set by being members of it. Leśniewski's term for the one thing is 'sum', and the individuals belong to the sum by being parts of it. The difference between these assumptions is not merely terminological, however. A Cantorian set is a whole conceived as a distributive class. A Leśniewskian sum is a whole conceived as a collective class.

In order to have a clearer picture of the difference between classical mereology and set theory, a few basic terms need to be explained. We may take the term 'proper part' as primitive. In saying that a is a 'part' of b (where a and b are individuals), we mean that a is either a proper part

---

[33]For details, see Machover 1996, 13ff.; Burton 1985, 613-615.

[34]See Simons 1987, p. 60.

of, or identical to, *b*. The relation indicated by 'proper part' is irreflexive and transitive. That is, if *a* is a proper part of *b* then *b* is not a proper part of *a*; and if *a* is a proper part of *b* and *b* is a proper part of *c*, then *a* is a proper part of *c*. Two individuals may 'overlap' one another, that is, they may have some part in common. If they do not overlap, they are said to be 'disjoint' from one another.[35]

Whenever there are two individuals, *a* and *b*, there is a unique individual described as the 'sum' of *a* and *b*, where *a* and *b* are parts of their sum. More generally, whenever there are many individuals, $a_1, \ldots a_n$, there is a unique sum of those individuals, where each of the many individuals is a part of their sum. Thus we can see that just as in set-theory a particular set can be defined by reference to a list of members, in classical mereology a particular sum can be defined by reference to a list of parts. Note, however, that whereas a set of individuals is not assumed to be an individual, a sum of individuals *is* assumed to be an individual.

The sum of two individuals *a* and *b* is understood as an individual which stands in a particular relation to them, a relation which we may designate as 'is a sum of'. This relation can be defined more specifically as follows: By saying that *c* is a sum of *a* and *b* we mean that anything which overlaps *c* overlaps either *a* or *b*, and vice versa. More generally, taking the variables $x_1, \ldots x_n$, *y*, and *u* to range over individuals, we define

D1 *u* is a sum of $x_1, \ldots x_n =_{def.}$
for all *y*, *y* overlaps *u* if and only if *y* overlaps one of the *x*s (i.e. at least one of them)

Given that the definition of a sum in classical mereology is meant to capture a certain common-sense notion of a whole, its intuitive plausibility is evident. In paraphrase, it tells us that the sum is something within the boundaries of which you will be, as it were, when (and only when) you are within the boundaries of one of the parts. This expresses our sense that a whole is something which includes or contains all of its parts, and nothing else. Considered in this way, the definition seems quite innocent and hardly controvertible. It would not be unreasonable to claim that someone who denies that a whole is related in this way to its parts has

[35] There are different ways of presenting the definitional connections between 'part', 'proper part' and 'overlap', according to which of the terms one chooses as primitive. My presentation here follows that found in Simons 1987, p.37.

simply failed to grasp what is meant by 'whole'.[36] However, classical mereology assumes something much stronger than the claim that wholes have this feature, that they bear the relation *is a sum of* to specified parts.[37] For it implies that being a sum of specified parts *is the same thing as* being composed by those parts. In other words, classical mereology takes it that

D2 $x_1, \ldots x_n$ compose $u =_{\text{def.}}$
u is a sum of $x_1, \ldots, x_n$

To define 'compose' in this way is to ignore the possibility that there is more to composition than a whole's being a sum of its specified parts. It might be the case that different relations between a whole and its parts exist, that there are different kinds of composition, where each of these relations entails – but is not entailed by – the whole's being a sum of those parts. Consider the following analogy: Suppose A, B and C are the first, second and third floors of a three storied building, and suppose it is proposed to describe the relation between C and A as the relation *is above*. A critic might object that there is more to the relation than C's merely being above A. B is equally above A, and yet C and B bear different relations, respectively, to A. Being two floors above A entails being above A, just as much as being one floor above A does. But neither is entailed by being above A.

A critique of classical mereology along these lines is exactly what we would expect from someone who argues for what I described in section II above as the Platonic approach, for this approach is distinguished by the suggestion that wholes need to be accounted for in terms of a three tiered-structure, and the closely associated suggestion that there are different kinds of composition. Of course, merely to point out the logical coherence of rejecting the classical mereological definition of 'compose' is not enough. In order for the critique to be taken seriously, we expect the critic to offer an alternative to this definition which spells out specifically what more there is to composition, which the definition fails to express. For the

---

[36] Van Inwagen describes the thesis that 'a whole is the sum of its parts' as an 'obviously trivial truth'. See van Inwagen 1990, 30.

[37] It will be convenient to use expressions in italics to designate relations and properties, e.g. the relations *is a sum of, is a part of, overlaps,* and the properties *is round, is green, is one, is many.*

time being, however, it is enough to note the aspect of classical mereology which a Platonic criticism will focus on, which will allow us to distinguishes this from other sorts of criticism of classical mereology.

Before turning to the other criticisms alluded to, however, let me complete my brief characterization of the contrast between classical mereology and set-theory, by presenting basic assumptions of classical mereology in comparison with the six basic assumptions of set-theory which I listed above. I shall discuss the assumptions of classical mereology in greater detail in the course the book's argument.

(M1) The Principle of Uniqueness of Sums: It has already been alluded that the sums of classical mereology comply with this principle, according to which if u is a sum of $x_1, \ldots, x_n$, and v is a sum of $x_1, \ldots, x_n$, then u is identical to v. It is clear that this principle is analogous to set-theory's Principle of Extensionality, and it allows us in an analogous way to identify a sum by specifying a list of individuals of which it is a sum. With this principle in mind, classical mereology is sometimes is also called 'classical extensional mereology'. As I noted in (S1), the wholes of common-sense need not comply with this principle.

(M2) The Principle of Universal Existence of Sums: Again, it has already been alluded that sums comply with this principle, which states that *whenever* we specify individuals, some individual exists which is a sum of those individuals (and which in particular has them as parts). Here the analogy with set-theory's Comprehension Principle is evident. The classical mereological principle, however, is weaker than the set-theoretical one: The parts may be specified either by listing them or by specifying a property which they have, but in the latter case only if in fact some individual has this property. Again, as noted in (S2), common sense does not acknowledge the existence of wholes composed by arbitrarily specified parts.

(M3) Transitivity of parthood: I noted above that *is a proper part of* is a transitive relation, and the definition of 'part' in terms of 'proper part' guarantees that *is a part* is similarly transitive. In view of (S3), classical mereology presents in this respect a notion of a whole closer to that of common sense than the one presented by set-theory.

(M4) One way of belonging to a sum: While according to set-theory something can belong to a set either as a member or as a subset, according to classical mereology something can belong to a sum in only one way, as a part (the distinction between part and proper part is analogous to the distinction between subset and proper-subset). In this respect classical mereology agrees with what we tend to assume regarding wholes.

(M5) No null individual: In classical mereology there is no analogue to set-theory's notion of a set without members. An individual cannot be a sum of *no* individuals. It might have no proper parts, in which case it is described as an 'atom'. However, such an atom will be a sum of some individuals, namely, a sum of itself. The null set belongs to every set (as one of its subsets). By contrast, classical mereology does not suppose that some individual belongs (in any way) to every individual. Therefore, in this respect too it is in better agreement than set-theory with what we tend to assume regarding wholes.

(M6) No distinct singleton: According to classical mereology, an individual cannot have a single proper part, i.e. a single part which is distinct from that individual. A singleton in set-theory is a set which has a single member, and such sets are distinct from their members. Thus classical mereology has no analogue for a set-theoretical singleton. As I noted in (S6), this corresponds with the way we tend to conceive of wholes.

We see, then, that classical mereology reflects some of our common-sense assumptions about wholes better than set-theory does, while conflicting with other common-sense assumptions.

In recent decades, the renewed interest of analytical philosophers in metaphysical questions, and particularly in questions about the nature of concrete entities and their identity through time, has given reason to clarify the notions of whole and part as they apply to such concrete entities. It will not be surprizing to find that the different views arrived at regarding wholes and parts depend to a significant degree on differences in the theorists' basic metaphysical assumptions and intuitions. Theorists sympathetic with the tendency commonly associated with modern physics, to treat the microscopic as ontologically and explanatorily prior, and to treat time as a fourth dimension along which concrete entities extend, have usually been attracted to the formal elegance and simplicity of classical mereology, and unimpressed with the ways in which it conflicts with pre-philosophical ways of using the terms 'whole' and 'part'. On the other hand, theorists less inclined to treat the substances of common sense, and especially living organisms as microphysically reducible, or to give up the idea that concrete entities can preserve their identity in spite of change, have often put forward claims about wholes and parts incompatible with classical mereology. The controversies mainly concern the principles of Uniqueness of Sums and of Universal Existence of Sums.[38]

I shall describe as a 'neoclassical mereology' any theory of wholes and parts which rejects either the Uniqueness of Sums or the Universal Existence of Sums (or both), but otherwise accepts some version of the assumptions of classical mereology. It seems to me that the principal contemporary positions regarding wholes and parts may be classified as versions of either classical or neoclassical mereology. The contrasts between these two broad families of views are interesting and important, and I shall look into them in some detail in what follows. However, I shall be primarily interested in what they have in common, and particularly in their shared assumption that the notion of composition is simply the flip side of the notion of being a sum (expressed in D2 above), whereby wholes are necessarily understood as two-tiered structures, in accordance with what I have called the Aristotelian approach. In view of their common core I shall describe them as sharing a *conception of wholes as sums*. I shall be discussing the weaknesses and strengths of this conception, and shall attempt to show how the weaknesses point to an alternative *conception of wholes as Unities*, whereby wholes are understood as three-tiered structures, in accordance with the Platonic approach. Recall that in such a structure we find many parts in the first tier, a lower-grade whole in the second tier, and a higher-grade whole in the third tier. The parts are related to the lower-grade whole by one sort of composition, similar in its characteristics to composition according to the conception of wholes as sums. They are related to the higher-grade whole (the Unity properly speaking) by a different sort of composition, the characteristics of which are incompatible with definition D2.

---

[38] For details, see Chapter 5 below.

V - Outline of the argument

The main body of this study is itself a whole, composed of three parts. However, it is primarily concerned with the metaphysics of concrete wholes, not abstract wholes like itself. An initial impression of the course of my argument can be gained by reviewing the titles of the different parts. In the first part, 'Wholes', I consider the notion of a concrete whole as it were pre-theoretically, that is, from a perspective prior to the acceptance of one or another fully developed mereology. In the next two parts, 'Sums' and 'Unities', I present contrasting theoretical accounts of wholes as sums and as Unities, respectively. The argument is, briefly, that the familiar conception of wholes as sums is, in spite of its flexibility, crucially unsatisfactory, and that the novel conception of wholes as Unities offers a promising alternative. Somewhat analogously, one might argue for a dualistic conception of mind by first discussing mental phenomena pre-theoretically, then proceeding to present and criticize a materialistic theory of mind, and finally formulating and demonstrating the explanatory advantages of a dualistic theory of mind.

Part One is devoted to exploring and articulating a more encompassing category within which the category of concrete wholes is to be located. Concrete wholes, I claim, are a kind of 'comprising entity', and by examining fundamental distinctions between different kinds of comprising entity I aim to make it seem less strange to suppose that there is more than one reasonable candidate for a theoretical account of wholes.

Comprising entities can be roughly explained as entities to which other entities – the comprised entities – are assumed to 'belong' in some way. A tree, a collection of stamps, the set of even integers, are all examples of comprising entities. The category includes concrete entities and abstract entities. The ontological status of some of them, especially those we tend to describe as concrete individuals, is less controversial than that of others. Some comprising entities are naturally described as wholes, and the corresponding comprised entities are described as parts. Indeed, it is arguable that if 'whole' is taken in a sufficiently wide sense, all comprising entities are wholes. However, if we are interested in the notion of a whole as this is understood in connection with concrete individuals, we have a narrower sense in mind. Assuming that we do, therefore, it seems that the relations between wholes and parts are a special case of relations between comprising and comprised entities. By clarifying the ways in which comprising entities might differ from one another, and particularly the ways in which the relations between comprising and comprised entities might differ from one another, we shall be in a better position to define concrete wholes as a species of comprising entities.

One of the most obvious aspects of relations between comprising and comprised entities is that in some sense or another they are relations between one and many. I distinguish between a number of senses of 'being one', the most important of which I describe as 'being unified', 'being monadic' and 'being single'. Very briefly, a comprising entity is unified if the entities it comprises are 'closely connected' to one another; it is monadic if it is distinct from the comprised entities; and it is single if it is appropriate to attribute to it predicates in the singular. This vocabulary will help us to draw fundamental distinctions between kinds of comprising entities. Such distinctions will be further facilitated by making use of the device of plural quantification. All of these preliminaries can be found in Chapter 2.

In Chapter 3 I discuss three kinds of comprising entity, collective classes, distributive classes, and collections. I argue that both a collective class and distributive class might be concrete, and that (in case they are one in the sense of singleness) they are kinds of collection, in a specific sense of the term. Collections are *coextensively determined* comprising entities. This feature will of course be explained in detail, but an expression of it is that the collection which comprises stones *a*, *b* and *c* (at some moment of time) cannot be identical to the collection which comprises stones *a*, *b*, *c* and *d* (at the same moment of time).

These observations lead me to suggest in Chapter 4, after explaining basic terms associated with the pre-theoretical notion of a whole, that three general models for a theory of wholes are available. I call them the conception of wholes as Classes (upper case C), as sums, and as Unities (upper case U), respectively. The first takes wholes to be (concrete) distributive classes. The second takes them to be collective classes. The third takes them to be a kind of comprising entity which, by contrast with collective and distributive classes, is not a collection (and thus is not coextensively determined).

It will be suggested in the second part of the book that classical mereology and neoclassical mereology are developments of the conception of wholes as sums, implying that current views by and large construe wholes as collections of some sort, in my sense of the term. If this is correct, then it seems that two further major options are largely neglected. The first, that of wholes as Classes, also takes wholes to be collections, though of a different sort, and it is a relatively straightforward matter to describe it in formal terms. The other option, which takes wholes to be Unities, however, is at this stage stated only negatively, in saying that Unities are *not* collections. In the present study I shall only be interested in developing this second option (in Part Three), as an alternative to the conception of wholes as sums (which I present and criticize in Part Two). It is worth emphasizing that, against what might be expected (from the perspective of an Aristotelian approach to wholes and parts), the differences between Classes, sums and Unities are made out without any reference to relations between the entities which they respectively comprise.

The usual technical term for a whole in connection with classical and neoclassical mereology is 'sum'. These theories do not allot distinct technical terms for a part and for the relation *compose*, but the terms 'part' and 'compose' as used in presenting the theories should be viewed, strictly speaking, as technical terms. In the case of the other two models, there is no established usage, and we are free to choose a convenient technical terminology. My choices are displayed in the following table:

Pre-theoretical	Classes	sums	Unities
'Whole'	'Class'	'sum'	'Unity'
'part'	'member'	'part'	'element'
'compose'	'form'	'compose'	'underlie'

The more detailed presentation of the various subdivisions of the category of comprising entities, in the text of Part One, will hopefully make them much clearer. In the mean time the reader may consult their representation in Figure 9.

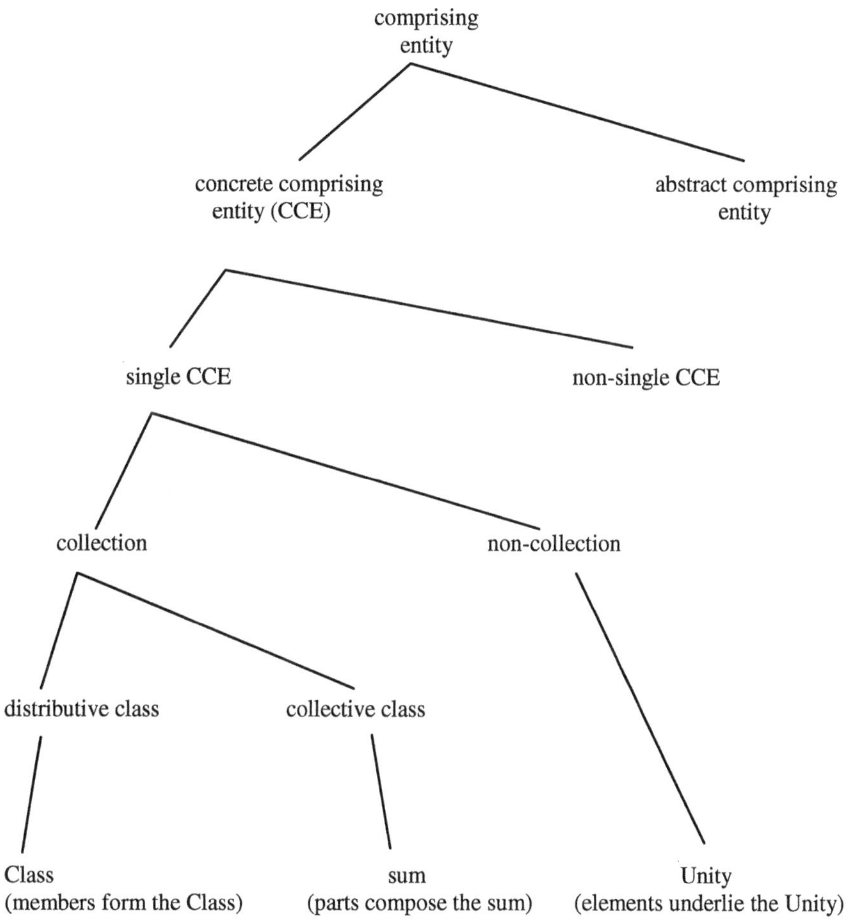

Figure 9

Part Two presents classical and neoclassical mereology as versions of the conception of wholes as sums, and attempts to point out some of the strengths and weaknesses of these theories. In chapter 5 I consider reasons which lead many theorists to prefer some form of neoclassical mereology to classical mereology – principally that they are thus enabled to describe 'conditioned sums' (sums the existence of which depends on there being certain relations between the parts), 'non-unique sums' (cases in which more than one sum is composed by the same parts at the same time), and 'mereologically varying sums' (sums which lose or gain parts over time). I argue that the theoretical modifications introduced by neoclassical mereology leave untouched the fundamental assumption that to be composed of parts is simply to be a sum of those parts.

It might supposed that if the conception of wholes as sums fails to capture a certain subtlety or richness in the nature of wholes, this should be revealed by comparing it with views associated with traditions which laid great emphasis on the (variously understood) claim that 'the whole is greater than the sum of its parts', and in particular with the ideas of an organic whole and a Gestalt. In Chapter 6 I argue to the contrary, that it is well within the resources of neoclassical mereology to describe organic wholes and Gestalts, at least to the extent that we consider these (often intuitively motivated) ideas to be fairly represented by the more explicit accounts which their proponents have provided.

A closer look at the way neoclassical mereology explains such phenomena as conditioned sums, non-unique sums, and mereologically varying sums, however, raises serious doubts as to the adequacy of the conception of wholes as sums. Chapter 7 develops a criticism of neoclassical mereology which focuses on these issues. In Chapter 8 I try to trace the difficulty back to the assumption implied by the conception of wholes as sums, that wholes are (what I earlier described as) coextensively determined. The source of the difficulties with sums can be seen from a different perspective in the plausibility of the claim for which I argue (against the prevalent view) that a classical sum is identical to its parts. Thus it is not *one*, in the sense that it is not monadic. If it is possible for wholes to be identical to their parts, monadicity becomes a non-trivial feature of the wholes that do possess it. And it becomes a test for a theory of wholes, whether it is successful in describing such wholes. Neoclassical mereology presents sums which apparently are monadic. However, its weaknesses may indicate that the conception of wholes as sums is ultimately irreconcilable with the notion of a truly monadic whole.

In Part Three the need for an alternative conception of wholes, to which the criticisms in Part Two have pointed, is met by proposing a theory of wholes as Unities, indicating ways in which it might be further developed, and demonstrating its power in treating problems which arise in the metaphysics of concrete wholes.

Chapter 9 presents the notion of a Unity, first by explaining it intuitively, and then by formulating a detailed systematic theory which is comparable to, and yet crucially distinct from, neoclassical mereology. In neoclassical mereology (and *a fortiori* in classical mereology), the relation *compose* is understood in such a way that if mutually disjoint entities *a*, *b*, and *c* compose the whole *d*, then necessarily the parts *a* and *b* do *not* compose the same whole at the same time. The theory of Unities, by contrast, distinguishes between two kinds of the relation *compose*, only one of which complies with this neoclassical constraint. A whole bearing this relation to the parts is described as a 'collection' of those entities. The second kind of relation, termed 'underlie', allows for *a*, *b*, and *c* , on the one hand, and *a* and *b*, on the other hand, to compose the same whole at the same time. The whole in this case is described a 'Unity' which the entities underlie. Whenever entities underlie a Unity, there is also at the same time a collection of those entities. Thus the Unity which *a*, *b* and *c* underlie is a whole associated with a three-tiered structure. The first is that of the (many) entities, *a*, *b* and *c*. The second is that of one whole, the lower-grade whole, which is a collection of those entities. The third is that of one whole, again, but of a higher-grade whole, the Unity which the entities underlie.

I argue that while a collection is identical to the entities of which it is the collection, a Unity is distinct from the entities which underlie it. Unities, therefore, are monadic, whereas collections are not. This is one expression of the clear sense in which Unities are wholes more loosely determined by their parts than collections or indeed sums. Unities are less sensitive to conditions at their boundaries than the other sorts of wholes are. This, I claim, makes it easier to explain such features of concrete wholes as the typical vagueness of their boundaries, and the ease with which they seem to survive loss or gain of parts.

In Chapter 10, the last, I try to illuminate the notion of a Unity from various perspectives, arguing that, among other things, it involves a conception of the way many parts belong to the same whole as somewhat analogous to the way many *properties* belong to the same substance. Thus, just as the identity of a substance may be fixed by a description which mentions only some of its properties, the identity of a whole may be fixed by a description which mentions only some of its parts (i.e. parts of which the whole is *not* a sum), as long as those parts compose (i.e. underlie) the whole. Furthermore I explore in a tentative manner ways in which assumptions can be added to the theory of Unties to facilitate a more complete account of concrete wholes.

Finally, I subject the theory of Unities to the telling test of confronting one of the best known and most intransigent paradoxes which afflict accounts of concrete wholes, often presented in terms of the example of Tibbles, the cat that loses its tail. My claim is that the theory of Unities offers an elegant solution to this paradox, avoiding the various implausible assumptions which other metaphysicians, concerned as I am with taking identity through time seriously, and yet working within the framework of the conception of wholes as sums, have needed to resort to.

Part I: Wholes

Chapter 2
Concrete Comprising Entities

I - Preliminaries

It is a familiar and pervasive feature of thought and speech that often when we are able to refer to many distinct entities, $x_1, x_2, \ldots, x_n$, we also seem able to refer to an entity y which in some loose sense we take to include or contain $x_1, x_2, \ldots, x_n$. Let us say that in such cases y *comprises* $x_1, x_2, \ldots, x_n$, and so describe y as a *comprising* entity. A tree, we may say then, comprises its cells. The set (in the ordinary, non-mathematical sense) whose members are the tree's cells may also be said to comprise those cells. We usually consider the relation between the cells and the tree to be *compose*. The relation between the cells and the set of cells we normally take to be a different relation, *are all the members of*. However, these relations have general features in common which seem to justify our taking them as species of one and the same generic relation.

In the first place, while the cells are *many* entities, the tree is *one* entity, and similarly the set of cells is *one* entity: Thus both relations are relations between many entities and one entity. Secondly, there is a sense in which the existence of both the tree and the set of cells depends necessarily on the existence of the cells. Suppose the cells did not exist, but that the world was otherwise exactly as it is, except in respect of everything which depends necessarily on the existence of the cells. Necessarily, in such a world, neither the tree nor the set of cells would exist.

Thirdly, with regard both to the tree and the set of cells, it seems to be an appropriate use of metaphor to say that the cells *belong* to them: The cells belong to the tree, and the cells belong to the set (though admittedly 'belong' does not have precisely the same sense in the two cases). Finally, it is at least *prima facie* plausible loosely to describe speaking of either the tree or the set by saying that in either case we speak of the many cells 'as one entity'.

We say therefore that both the tree and the set are *comprising entities*, and that the cells are comprised by both the tree and the set. By contrast, it is clear that the relations between David and his many friends and between David and his many books are not relations between comprising and comprised entities, although both are relations between one entity and many entities, and the verb 'to belong' is used to describe the relation between the many entities and the one entity. For the second and fourth feature are missing in these cases.

Ultimately, however, the decision to speak about a generic comprising relation need not be treated as more than a heuristic device, a way of focusing on several very general problems without getting bogged down at this early stage of the discussion in the numerous terminological distinctions which abound in this field. With this in mind, it seems best to avoid attempting a precise characterisation of the comprising relation, and instead to have before us typical specific examples.

Consider therefore the following table, in which expressions designating many comprised entities are listed in the left-hand column, and expressions designating corresponding comprising entities are listed in the right-hand column:

COMPRISED ENTITIES	COMPRISING ENTITY
stones	wall
molecules	stone
molecules	tree
cells	organism
cells	collection of cells
pebbles	collection of pebbles
pebbles	class of pebbles
pebbles	set of pebbles
birds	flock
players	team
grains of sand	heap
buildings	group of buildings
stars	star-cluster
meteorites	collection of meteorites

The expressions listed in the left hand column are in the plural, syntactically speaking: They combine with verbs in the plural to form grammatical sentences. Thus on syntactical evidence each of them is apparently used to refer to many entities. The expressions listed in the right hand column are syntactically either singular or indeterminate (with respect to number). 'Organism' combines with verbs in the singular to form grammatical sentences, while 'team' combines either with verbs in the singular or with verbs in the plural to form grammatical sentences ('The organism survives in harsh conditions'; 'The team plays/play well').

On syntactical evidence, again, each of the singular expressions is apparently used to refer to one entity, as are each of the indeterminate expressions in contexts where they combine with verbs in the singular. I shall assume in what follows, unless explicitly stated otherwise, that indeterminate expressions are found in such contexts, and thus I shall treat them as syntactically singular. I shall take it then that each of the expressions in the left hand column is apparently used to refer to many entities, whereas each of the expressions in the left hand column is apparently used to refer to one entity.

One point should be emphasized about my use of 'comprises': I say that the organism comprises the cells; I do not say that the organism comprises *each* of the cells. 'Comprises' expresses a certain relation which holds between the organism, on the one hand, and the cells taken together, on the other hand.

Regarding some of the examples in the table the existence of a comprising entity is not controversial, or at least no more controversial than the existence of the comprised entities. It is normally assumed for instance that there exists an entity to which 'organism' is appropriately used to refer, no less than it is assumed that there exist entities referred to by means of the expression 'cells'. What *is* controversial here is how we are to conceive of the relations between the comprising and comprised entities. Pre-theoretically, it is a relation between a whole and the many parts which compose it. Various theoretical accounts have been offered for this relation, however, and each has its own difficulties. We will be looking more closely at the theory of wholes and parts in later chapters.

Regarding some of the other examples in the table, however, the *existence* of a comprising entity is much closer to the foreground of controversy. Common sense does not seem to have as confident a position as regards the existence of entities to which expressions such as 'team of players', 'flock of birds', 'collection of pebbles' or 'set of pebbles' might be used to refer, as it did regarding the existence of a wall, a stone, a tree, or organisms in general.

Doubts of this sort arise in particular in connection with many terms used in more or less narrow technical senses by various authors in their discussion of comprising entities. Thus, for example, Russell and Whitehead argue in *Principia Mathematica* that classes are logical fictions.[1] And in his *Introduction to Mathematical Philosophy* Russell expresses scepticism regarding the items designated by 'heap' or 'conglomeration'.[2] Among these terms we find, in addition to 'whole', 'sum', 'class', 'set', 'collection' (of which I shall be making much use), the following: 'fusion' (see Lewis 1991, 72-4); 'manifold', 'domain', aggregation (see Frege 1895, 86-7); 'heap', 'aggregate';[3] 'conglomeration' (see Russell 1919, 183); 'group' (see McTaggart 1921, Ch.XV; Sprigge 1970, 112ff.); 'totality' (see Küng 1967, 105-8). Although I take the notions associated with each of these terms to be notions of comprising entities, I do not attempt in the present work to settle their precise relations to notions that occur in my discussion.

The sort of reasons which might lead us to doubt the existence of entities intended by such terms can be indicated by considering an example. Suppose the expression 'collection of pebbles' is used in a context in which more than one pebble is assumed to be a member of the collection. It might be suggested that the sentence 'the collection of pebbles is being moved' means the same as 'the pebbles are being moved'; and that therefore it is inappropriate to acknowledge in this context the existence of any entities to which the latter sentence does not commit us. But the latter sentence commits us only to the existence of the many pebbles, and the collection of pebbles, if there were such a thing, could not be identical to any of the pebbles. Alternatively, it might be suggested - if it is denied that the two sentences have the same meaning -

[1] Russell and Whitehead 1910; See also Goodman's (1956, 199-203) criticism of the notion of a class.

[2] Russell 1919, 183.

[3] These two terms seem to be treated as synonymous by Armstrong 1978, 29ff.. His comments on p.31 suggest that he understands them in Goodman's sense of 'sum' (or 'sum-individual' - see 1956, 201). Quine uses 'heap' in this sense in 1953, 114. For an earlier use of 'aggregate', see Russell 1903, 138-140. Both terms are common in commentary on Aristotle, as translations of the Greek '*sōros*'. Miller (1978, 112) argues that in some contexts, though not all, Aristotle uses '*sōros*' in a technical sense. The translation 'aggregate' seems to emphasize the technical sense - see particularly the account of the notion in Scaltsas 1994b, 66, and 1994a, 111. For further examples of the use of 'heap' and/or 'aggregate' in commentary on Aristotle, see e.g. Furth 1988, 294 (index item); Haslanger 1994, 140; Halper 1989, 139, 151; Bogaard 1979, 12.

that the notion of a collection is obscure, and that the existence of an entity corresponding to such an obscure notion must be doubted.[4]

In the case of a doubtful comprising entity the question whether the entity exists is closely bound up with the question how such an entity should be conceived. For example, if one takes it that the notion of a collection must be construed as the notion of a certain type of abstract entity, then if one denies the existence of abstract entities one will also deny the existence of a collection. One might then say that notwithstanding the linguistic evidence according to which reference is apparently made to a comprising entity, analysis reveals that this reference is only apparent, and no comprising entity exists which would correspond to the expression 'the collection' in the intended contexts.

Questions about the way a doubtful comprising entity should be conceived are closely connected, furthermore, with questions about the relations between the comprising and comprised entities. If for example one speaks of a collection of cells one is likely to be led much more directly to a consideration of the relation between the ostensible comprising entity and the individuals it comprises, than one would if one spoke of an organism. For one is more likely to be faced immediately with clarificatory questions such as: When you speak about the collection of cells, do you mean simply to speak about the many cells, or do you mean to distinguish between talk about the collection and talk about the many cells? Of course such questions can be asked in connection with the organism no less than they can with regard to the collection of cells. But in the case of the collection they seem, as it were, closer to the surface, and more urgently in need of an answer for any discussion of the ostensible entity to proceed with clarity.

Taking these points into account, one is less surprised by the realization that the principal current theories which attempt to account for the relations between *uncontroversial* comprising entities and the individuals they comprise (e.g. the organism and the many cells) are descendants of 19th century theories which were principally concerned with *controversial* comprising entities (e.g. classes, sets, manifolds). If one wishes to examine fundamental features of current theories of wholes and parts, therefore, it seems reasonable to begin by concentrating on conceptions of ontologically more controversial or doubtful comprising

[4] See Max Black's claims regarding the obscurity of the notion of a set, in Black 1971; see also his comments on the notions of a collection and an assembly, p. 619-22.

entities. This is an important reason for the emphasis laid in Part One of this book on a discussion of the latter type of comprising entities.

Ultimately, however, the book's main concern is to examine the conception of *uncontroversial* comprising entities. I take the following to be such entities: Organisms (such as trees, whales, viruses), more or less clearly delimited chunks of matter (such as rocks, mountains, lakes, planets), artefacts (such as houses, guitars), and various parts of any of these (such as cells, kidneys, molecules). These are all concrete entities, and they are all, at least *prima facie*, individuals. Moreover, they are entities which comprise concrete entities (i.e. their parts) which are *prima facie* individuals. Our investigation does not require a precise delineation of the range of entities with which we will be concerned. However, since it has been described as the range of concrete entities which are *prima facie* individuals, a brief explanation is called for of the way I understand the notions of concreteness and of an individual.

1. Concreteness

For the purpose of indicating the sort of comprising entities I am ultimately interested in here, I speak of 'concrete entities' in a relatively narrow sense of the expression.[5] An entity is concrete if (a) there is a more or less definite interval of time during which it exists as a *continuant*, (b) there is a more or less definite region of space which it occupies at any moment of its existence, and (c) there is a more or less definite mass (or weight) which it possesses.

The distinction implicit in (1), between continuants and occurrents, refers to a contrast between two ways in which entities can exist in time, and thus it is clear in particular that concrete entities exist in time. Intuitively speaking, continuants *endure* through time, whereas occurrents *extend* through time. I offer the following fuller explanation of this distinction, taking the predicate 'exists during *I*' (where *I* is a temporal interval) to apply equally to both kinds of entity, and to be primitively understood.[6] We may define 'exists at *t*' in terms of 'exists during *I*' as follows:

x exists at $t =_{\text{def.}}$

[5] For a discussion of difficulties in defining concreteness more generally, see Hale 1987, ch.3.

[6] For different accounts of the distinction between continuants and occurrents, see Simons 1987, 132; Mellor 1995, 123.

for some I, such that I is an infinitesimal temporal interval
and t is a moment of that interval,
x exists during I.

The terms 'continuant' and 'occurrent' can be defined as follows:

x is a continuant $=_{\text{def.}}$
x exists during an interval I
if and only if for all t, such that t belongs to I, x exists at t.

x is an occurrent$=_{\text{def.}}$
x exists during an interval I
if and only if
(1) for all t, such that t belongs to I, and for some y,
y is a part of x and y exists at t.
and
(2) for all t, t', such that both t and t' belong to I, but $t \neq t'$,
and for all y, z, such that both y and z are parts of x,
and y exists at t and z exists at t',
$y \neq z$.

Given this explanation of a continuant's existence *at* every moment of the interval *during which* it exists, the meaning of second condition (b) is well defined in a way it would not be if it were applied to occurrents (I have not assumed that an occurrent exists *at* every moment of the interval *during which* it exists). As for condition (c), it is added in order to exclude such entities as souls and gods, as well as shadows, images and the like. The burden of the qualification '*prima facie*', when I speak of 'entities which are *prima facie* concrete', is born mainly by condition (a). An entity is *prima facie* concrete if it is, *prima facie*, a continuant, it being controversial whether construing a horse, for example, as a continuant is appropriate from a theoretical standpoint.

2. Individuals

As regards the notion of an individual, its uncontroversial core seems to involve (d) the idea of an entity which is in some sense separate or independent from other entities, and (e) the idea of an entity which is in some sense *one*. Both of these ideas are ambiguous, vague, and to some extent obscure.

Concerning (d) I shall only add that it is commonly understood as implying that properties, relations, and sets or classes are *not* individuals, because they are not separate or independent from other entities in the required way. I shall simply assume that properties, relations and sets are not among *prima facie* individuals, without attempting here to clarify the sense in which the notion of an individual involves separateness or independence.

As for (e), it is ambiguous because the expression 'is one' is associated with a variety of senses, and it is obscure to the extent that we are unable to distinguish between and explicate these senses satisfactorily. It seems reasonable to suppose that every comprising entity, indeed, anything properly described as *an* entity, is one in some sense or another. In the next section of this chapter I shall offer my own account of the range of senses of the expression. In the mean time let me note that if two distinct senses A and B are distinguished, and a comprising entity may be one in sense A without being one in sense B, then we must decide which sense of 'is one' is relevant as far as the entity's being an individual is concerned. Different theories may differ from one another, and from pre-theoretical thinking, in their view as to the relevant sense of 'is one'.

It seems clear that entities which are *prima facie* concrete individuals have a feature which may be described as 'unity', providing us with one of the senses of 'is one' which I shall be mentioning in the survey of the next section.[7] It is useful to present it here, however, because it will help to dispel some of the confusion associated with the notion of an individual.

It is not obviously the case that if an entity is one, it possesses the feature of unity. By saying that a comprising entity possesses unity or is unified I shall mean that the entities it comprises are closely *held together* – itself a rather vague notion which has, nevertheless, quite a strong intuitive appeal. Compare for instance (i) a whale with (ii) a *pair* of whales, an entity which comprises the bodily organs of two whales. The whale is *prima facie* an individual, the pair of whales is not. In the former case, the comprised entities seem to be closely held together in a way that they do not seem to be in the latter case. One expression of the contrast

[7] The use of 'unity' here, as expressing a *feature* of entities, should not be confused with the use I introduce later, where 'unity' is used to describe a *type* of entities. I distinguish the latter use distinguish from the former by employing the upper case 'U'. Thus and entity may *have* unity, but it may *be* a Unity. Note in particular that an entity which has unity need not be a Unity.

between these examples is that the organs of the single whale meet conditions of spatial continuity, which those comprised by the pair of whales do not. Another expression of the contrast is that the well-being of one of the organs of the single whale may influence the well-being of all the other comprised organs; the organs comprised by the pair of whales are not similarly interdependent.

Let us say that when comprised entities are closely held together, in this sense, there subsist between these entities relations of 'integrity', or equivalently that they are 'integrally' related to one another.[8] Thus a comprising entity is unified if and only if the entities it comprises are integrally related to one another. Of course, it is a further question, which relations should be viewed as relations of integrity, but settling this question is not necessary for our present purposes.[9]

Now, theorists may be compared in broad terms (overlooking various subtleties) with regard to their views of individuals by comparing their attitudes to the assumptions I call 'Unified Entities' (UE) and 'Unified Individuals' (UI), stated as follows:

UE An entity (i.e. a comprising entity) exists only if it is unified

UI An individual exists only if it is unified

Assuming that individuals are comprising entities, three possible views may now be distinguished:

(1) UE and UI
(2) not-UE and UI
(3) not-UE and not-UI

The typical upholder of (1) assumes that if entities $x_1, \ldots, x_n$ are not integrally related to one another, there is no entity y which comprises them. If they are integrally related to one another then some y comprises them, and y is an individual. Peter van Inwagen exemplifies this view, in assuming that only entities which collectively display the feature of 'life' are comprised by an entity, the individual organism.[10]

---

[8] Simons' uses 'integrity' in my sense of 'unity'. See Simons 1987, 326ff.
[9] Van Inwagen 1990 is largely concerned with this question.
[10] See van Inwagen 1990, 82.

The typical upholder of (2) assumes that for *any* entities $x_1, \ldots, x_n$, there is an entity y which comprises them. Only if the xs are integrally related to one another, however, will y be an individual. Thus, such a theorist acknowledges the existence of non-individuals as well as individuals. Peter Simons' view belongs here, in taking unity to be constitutive of being an individual, and yet acknowledging the existence of non-individual sums and groups.[11]

Finally, the typical upholder of (3) assumes like that of (2) that for any entities $x_1, \ldots, x_n$, there is an entity y which comprises them. By contrast, however, she will take it that in every such case y will be an individual. Nelson Goodman is an example of such a theorist,[12] and indeed this view is typical of those who consider classical mereology to provide an adequate account of concrete wholes.[13]

My own argument is best served by assuming a version of (2). I shall be discussing both un-unified and unified comprising entities, though I shall qualify the former as 'controversial'. At this point it is unimportant for my purposes either to affirm or to deny that un-unified entities are individuals. Ultimately, however, I will be presenting a theory which distinguishes between sums and Unities as two kinds of wholes, where the existence of a Unity does depend on its being unified. I am strongly inclined to interpret Unities as individuals, and therefore to assume in the manner of (2) that not all concrete comprising entities are individuals. Incidentally, I shall also be arguing that a comprising entity which is not a Unity is identical to its parts, and thus there is an important sense in which recognizing the existence of non-individuals will turn out to be ontologically harmless.

At any rate, at present I am assuming a distinction between controversial and uncontroversial comprising entities, taking the latter to be comprising entities which are *prima facie* continuants; which occupy a more or less definite region of space at each moment of their existence; which have a more or less definite mass; which are neither properties, nor relations, nor sets or classes in a technical sense; and which are *prima facie* unified. When using the term 'entity' I should be understood as speaking of either controversial or uncontroversial comprising entities, unless otherwise pointed out.

---

[11] See Simons 1987, 144-148; 290-293; 324-326.

[12] See Goodman 1956

[13] Lewis 1986a, 36-42, argues for this view of classical mereology. See also his 1991, 79-81.

The range of comprising entities I shall be considering, however, is still only partial. My attention will be confined to entities of which it makes sense to ask, how many of them there are (in a certain place, or answering a certain condition). It makes sense to ask how many pebbles there are on the beach. And it makes sense to ask how many *collections* of pebbles there are on the beach (although this second question may admit of several incompatible interpretations, which correspond to different assumptions that might be adopted concerning the proper way to count collections). But it does not make sense to ask 'how many water there is in the sea', or (unless one takes 'water' in the sense of 'a body of water') 'how many waters there are in the sea'. Rather, one may ask how *much* water there is in the sea.

This means that I shall be ignoring the ontological category of stuffs or masses. To the extent that I will have the occasion to mention such things as lumps of gold or heaps of sand, I shall be treating them under the tentative assumption that theories devised with organisms, pebbles, and collections of pebbles in mind are applicable to such lumps or heaps (considered, perhaps, as collections of molecules or collections of sand-grains). This is not because it is my considered opinion that stuffs do not require separate treatment, but rather because it would unacceptably complicate and enlarge this work, if such separate treatment were attempted here. I do not believe that the validity of my main argument will be affected by this simplifying assumption. It seems to me that bringing considerations to bear regarding the characteristics of stuffs will at most have the effect of restricting the generality of my conclusions.

II - Ways of Being One

The notions of a collection, a whole, a sum and a Unity are at the centre of the present work's concerns. They can all be described as notions of comprising entities. They will be introduced in turn, and the relations between them will be clarified as we proceed. One fundamental component of such a clarification needs to be treated already at this stage, however, before we turn to the specific discussion of any of these notions.

This concerns the ambiguity which attaches to the claim that a comprising entity is 'one', or 'one entity'.

Authors, both ancient and modern, have felt the contrary pull towards describing comprising entities (of some types, at least) as being both 'one' and 'many' (or 'not one').[14] The terms seem, initially at least, to be contraries, as the practice of counting suggests. It is evidently absurd to count horses and to report the result as 'both one and many'. Nevertheless we find many philosophers insisting in certain circumstances that both 'one' and 'many' apply to the same subject matter.

Plato, for example, explicitly emphasizes in one context that a whole is one, and in another context that it is many. On the one hand he argues in *Parmenides* 137c4-d3 that if something is a whole it consists of parts and 'is in that way many and not one' (and on these grounds he argues that the one itself 'cannot have any parts or be a whole'). Furthermore, in *Theaetetus* 204a7-205a9 he argues that 'when a thing has parts, the whole is necessarily all the parts', and so presumably it is many, since the parts are many. On the other hand, he describes the whole in *Parmenides* 157d8-e1 as 'some single form and one thing' and as 'a complete one which has come to be out of all [the parts]'.[15] And he makes a similar point in *Sophist* 245a1-3.

Perhaps less explicitly, what Aristotle says about heaps seems to suggest that he takes a heap to be both one and many. On the one hand, he argues in *Metaphysics* 1040b5-9 regarding 'the parts of animals ... and earth and fire and air' that 'none of them is one, but they are like a heap before it is fused by heat and some one thing is made out of the bits'.[16] Thus it seems that in some way the heap (before it is fused) is not one. On the other hand he appears to concede in 1015b36-1016a3 that things which are continuous, such as the parts of the body, are legitimately called one. So it would seem that in another way the heap - provided that it is continuous - is one.

It might be suggested that this apparent tension in what Aristotle says about heaps can be resolved by assuming that Aristotle uses the term 'heap' in some contexts in an ordinary sense (according to which the term may apply to continuous entities), while in other contexts in a distinct

---

[14] I shall be using the term 'many' in the sense of 'more than one'.

[15] The translation provided here of the last two quoted clauses is more literal than Cornford's 'a single entity or 'one' ' and 'a complete 'one' composed of all'.

[16] See also *Metaphysics* 1045a8-10.

technical sense.[17] However that may be, it is noteworthy that when Theodore Scaltsas presents a modern version of Aristotle's account, he affirms the tension between an aggregate's being one and being many, in a way which cannot be resolved by referring to such a distinction between senses of 'heap' or 'aggregate'.[18] For he says that 'the aggregate is not a single entity; it is the many things that it is the aggregate of'. And yet he claims that '...any number of things can be unified into a group just by including them in a group... . The ground for such unity is convention, and the resulting oneness is entirely compatible with the plurality that the things in the group constitute'.[19]

David Lewis's account of mereological fusions seems to suggest that he takes a fusion to be one, in one respect, but many, in another. He describes the relation of the many parts to their fusion as a 'many-one relation', apparently implying that the fusion is one. A few pages further along, however, he says that 'mereology is innocent in a different way [i.e. different from the way plural quantification is 'innocent']: we have many things, we do mention one thing that is the many taken together, but this one thing is nothing different from the many'.[20]

Peter Simons explicitly states that the sort of entity he calls a 'plurality' is both one and many: '... a plurality is in one sense one, and in another sense many. It is one class (as distinct from other classes), but at the same time it is (i.e. is composed of, has as members) many individuals.'[21]

In *The Principles of Mathematics* Bertrand Russell urges the notion of what he calls 'the class as many', and by contrast with the types of positions just mentioned insists with regard to such a class that 'the many are only many, and not also one'.[22] However, more than one commentator has found Russell's position here unintelligible. Max Black, for instance, takes Russell's account of the distinction between class as many and class as one to be an example 'of what Berkeley called the 'darkness and confusion of mathematics' '.[23] And it may be noted that Russell himself

---

[17] For example, Miller 1978, 108-9, claims that such a distinction between senses should be drawn.

[18] See note 3 above, regarding the use of either 'heap' or 'aggregate' to translate the Greek '*sōros*'.

[19] See Scaltsas 1994b, 66; 1994a, 109.

[20] See Lewis 1991, 82; 87. Lewis's position will be discussed in detail in Chapter 8 below.

[21] Simons 1987, 151-2.

[22] See Russell 1903, 76.

[23] Black 1971, 620-1; see also Grossmann's 1983 discussion of classes.

points out in the chapter 'The Contradiction' that in a certain respect the class as many is, after all, one.[24]

To observe such claims is not, of course, to imply that these well respected authors are all guilty of inconsistency. Examining their positions, one indeed finds different grounds for claiming that an entity is one rather than many. I shall consider features which provide such grounds as respects in which an entity is said to be one. Thus one finds that different respects may be distinguished in which an entity might be said to be one rather than many. Before turning to the more important of these respects, let us set aside other ways in which things are said to be one, which are of less interest in the context of a comparative characterization of comprising entities.

1. Linguistic Oneness

If x is a horse, then clearly x is *one* horse. It might then be said that x is *one*, where this is meant simply as a way of saying that x is one horse (a less awkward way of expressing x's being one in this respect would be to say that it is one *in number*). In this sense, a house is one, a pebble is one, a collection (say, of pebbles) is one, a heap is one, etc. In general, if 'F' is a term such that the sentence 'x is an F' is a (correct) description of x, then it is the case that x is one F; and so, in the sense we are considering, it is the case that x is one. Since expressions used to refer to comprising entities, such as 'whole', 'heap', 'sum', 'class', 'collection', etc., are all expressions of this type, it follows that whenever one of these terms is correctly predicable of x, then x is one in this sense.

For x to be one in this sense seems to be simply a consequence of the linguistic fact that a descriptive term (possibly, but not necessarily, a count-noun) in the singular is correctly predicable of x. To say that x is one in this sense is not so much to say something about the nature of x, as it is to say something about the way x is treated according to the conventions of the English language. Therefore, I call the feature of being one in this sense 'linguistic oneness'.

Frege's discussion in *The Foundations of Arithmetic* suggests that something's being one in number (compatibly with its being, say, five in number) is to be explained with reference to the fact that a certain term in the singular (e.g. 'copse') is correctly applicable to it:

---

[24] Russell 1903.

While looking at one and the same external phenomenon, I can say with equal truth both 'It is a copse' and 'It is five trees', or both 'Here are four companies' and 'Here are 500 men'. Now, what changes here from one judgement to the other is neither any individual object, nor the whole, the agglomeration of them, but rather my terminology. But that is itself only a sign that one concept has been substituted for another.[25]

Correlatively with the notion of linguistic oneness, a notion of *linguistic manyness* may be distinguished. If *x* and *y* are horses, then clearly *x* and *y* are *many* horses. We might say that *x* and *y* are many, meaning by this that for some term '*F*' the sentence '*x* and *y* are *F*s' is a correct description of *x* and *y*. We do not assume that being linguistically one is incompatible with being linguistically many. For example, if it is supposed that *z*, a sum of sand grains, is identical to two distinct sums of sand grains *x* and *y* (taken together), then since *z* is linguistically one, and *x* and *y* are linguistically many, and *z* is identical to *x* and *y*, it follows that *z* is both linguistically one and linguistically many.

Note that for *x* to be linguistically one does not imply that a *syntactically singular* term is predicable of it. Simons notes a threefold distinction between (1) syntactically singular, (2) syntactically plural, and (3) syntactically indeterminate terms.[26] This distinction has to do with the form of *verbs* combining with such terms to form grammatical sentences. The criterion is whether verbs combining with the term in question are singular, plural, or either singular or plural. Thus 'Socrates' is syntactically singular because we say 'Socrates *is* singing'; 'Lennon and McCartney' is syntactically plural because we say 'Lennon and McCartney *are* philosophizing'; and 'the orchestra' is syntactically indeterminate, because we may say either 'the orchestra *is* playing' or 'the orchestra *are* playing'. If *x* is an orchestra, then it need not be the case that a syntactically singular term is predicable of it (we have just noted that 'orchestra', for example, is not syntactically singular, but rather syntactically indeterminate). However, if *x* is an orchestra, then *x* is one orchestra, and so it follows that *x* is linguistically one.

---

[25] Frege 1884, 59.

[26] See Simons 1987, 142-3. The examples I mention in this paragraph are Simons'.

2. Atomicity

The claim that an entity is atomic is usually understood as the claim that it has no proper parts.[27] Since our discussion has not yet been confined to wholes and parts, and we are speaking of comprising entities more generally, it is useful to speak of a correspondingly more general notion of being atomic. Therefore, in the present context I take it that an entity is atomic if and only if it either comprises no entities at all, or it comprises only itself. The null class, for example, is atomic in this sense. And if a concrete entity has no proper parts then (assuming it comprises no entities which are not among its parts) it too is atomic in this sense. We see then that the more general notion of being atomic entails the narrower and more familiar one mentioned at the beginning of this paragraph.

Sometimes an entity will be said to be one on the grounds that it is atomic. Indeed, it seems natural to treat entities that are one in this respect as best deserving to be described as one, and thus as paradigms of being one. It seems that for this reason Plato claims that what he calls 'the one itself' cannot have parts, and that by having parts a thing is rendered (in some respect) many rather than one.[28]

Regarding the types of entities with which the present work is concerned, however, all the competing views which I consider agree in treating such entities as being *not* atomic, in the general sense I have noted. The notion of atomicity will therefore be of marginal importance in our discussion.[29]

3. Relational Oneness

Sometimes 'one' is used not in the context of sentences of the form 'x is one' but rather in the context of sentences of forms such as 'x and y are one' (or more generally 'x_1 , x_2, ... , x_n are one'). The notion of being one is understood in the latter type of contexts as that of a *relation* between entities.[30] In most cases of this sort, to say that x and y (or, more generally, x_1 , x_2, ... , x_n) are one is to say either of the following things: (1) x and y

[27] See, for example, Simons 1987, 16. I prefer the variant 'is atomic' to Simons' 'is an atom' here.

[28] See *Parmenides* 137c-d. That a thing which has parts can be one only in a more qualified way (than a thing which does not have parts is one) is also claimed in *Sophist* 245a-b. Note that, quite clearly, Plato uses 'part' in our sense of 'proper part'.

[29] The term 'atomicity' is sometimes used to describe the feature of an entity whereby it comprises parts which are atomic. I use the term differently, to designate the feature of being atomic itself.

[30] The relation might either be dyadic, polyadic, or multigrade.

are the same (either numerically, or in some particular respect), or (2) *x* and *y* compose (or, more generally, are comprised by) one entity, or (3) *x* and *y* are integrally related to one another. The notion of being one, as associated with either (1), (2) or (3), can be generically described as the notion of 'relational oneness'.

I note this notion of relational oneness here in order for it not to be confused with notions associated with use of the *monadic* predicate 'is one'. Any claim whereby a comprising entity is said to be one is a claim which utilizes the latter monadic predicate, and it is with such claims that we are now concerned. This is not to say, however, that some of the notions of being one associated with use of the monadic 'is one' may not be importantly bound up with the notion of relational oneness. This will be particularly evident with regard to the notion of unity, to which I now turn.

4. Unity

Perhaps the most common way to understand the claim that an entity is one, is to take it to mean that the entity is unified, that is, possesses the feature of unity, which as we have seen is taken by some authors to be constitutive of being an individual. It was explained that for a comprising entity to possess unity is for it to comprise entities which are integrally related to one another. It is clear that the notion of unity is closely connected with the notion of relational oneness. For a comprising entity is one, in the sense of being unified, if and only if the entities it comprises are one, in the sense associated with relational oneness.

An understanding of the notion of being one in terms of the notion of unity is suggested, for example, by van Inwagen's discussion of relations of integrity in 'When are Objects Parts?'. He is inquiring about which changes in the relations between many entities might be responsible for there coming to be an entity - one entity - which comprises the former. He seems to be assuming that one (comprising) entity comes into being when, and only when, a unified entity comes into being. What remains to be settled is the question, precisely which relations should be considered as relations of integrity, relations which constitute the unity of the new entity. He asks, for example, whether when 'you and I shake hands ... Does a new thing at that moment come into existence' (and answers in the negative, regarding this and other instances of the relation of contact).[31]

[31] See van Inwagen 1987, 28.

A similar sort of understanding of the notion of being one is found in Simons' discussion of relations of integrity, when he interprets Aristotle's claim that some entities 'are more truly one than others' in terms of the greater unity (which Simons calls 'integrity') of those entities.[32] He takes it that if a comprising entity is not unified, it will not be an individual but rather a collection, a kind of plurality, where a plurality is something he considers to be both one and many, as I noted above.[33]

It should be noted that talk about oneness that comes in *degrees* is most reasonably interpreted as talk about unity. It is natural to compare ways in which entities are related by saying that some are *more closely related* to one another than others. Relations of integrity might in this way be loosely taken as being of different degrees, so that those of a higher degree lend a higher degree of unity to the associated comprising entity. I shall not attempt to assess the appropriateness of such talk of degrees of unity.

5. Monadicity

In *Theaetetus* 203c4ff., Plato (as we saw in Chapter 1) contrasts the notion of a comprising entity which is identical to the entities it comprises (taken together), with that of a comprising entity which is not identical to the entities it comprises.[34] Describing the second type of comprising entity, he emphasizes a sense in which it is one by saying that it is 'some single form, produced out of them [i.e. the entities it comprises], having its own single nature – something different from the letters'.[35]

This suggests a further feature with reference to which an entity might be described as one: An entity y is one, in this sense, if and only if for all $n \geq 2$, and for all $x_1, x_2, \ldots x_n$ (all distinct from one another), y is not identical to $x_1, x_2, \ldots x_n$ (taken together). In particular, if y comprises the xs, and y is one in this sense, then it is not identical to the xs.

---

[32] See Simons 1987, 326.

[33] For his view of connections between unity (in his terms 'integrity') and the distinction between individuals and collections, see Simons 1987, 331; for the relation between 'collection' and 'plurality', see Simons 1987, 144-146.

[34] The claim that a comprising entity b is identical to comprised entities $a_1, \ldots, a_n$, is interpreted as the claim that the expressions 'b' and 'a_1 and ... and a_n' have the same reference.

[35] *Theaetetus* 203e3-5. The word 'single' appearing here in the Burnyeat (Levett) 1990 translation renders first '*hen*', then '*mia*', the normal masculine and feminine forms for 'one' in Greek.

I shall describe an entity which possesses this feature as 'monadic'.[36] Plato's contrast between two conceptions of a comprising entity in the *Theaetetus* is thus a contrast between monadic and non-monadic conceptions. Anyone who takes it, as do Doepke's One-Many theorists, and as Simons does, that one-many identity claims are improper and thus unacceptable in principle, will take it that *all* comprising entities are monadic.[37] I shall be arguing at a later stage, however, not only that such claims are acceptable in principle, but also that the sums of classical mereology are indeed non-monadic. I shall further argue that it is questionable whether theories which diverge from classical mereology, and yet retain fundamental features of the notion of a sum in their conception of wholes, have been successful in elaborating a coherent model for monadic concrete comprising entities. The notion of a Unity, which I introduce in Part Three of the present work, seems to offer a more promising model for monadic comprising entities.

It should be noted that there would seem to be no definitional connections between the notions of monadicity and unity which would allow the reduction of one of the notions to the other. The question whether an entity is monadic seems to be independent of the question whether it is unified. The impression that they are independent is encouraged by the fact that when Lewis considers the question whether a sum is identical to its parts (that is, whether a sum is non-monadic),[38] the issue of relations between the parts does not enter into his considerations. And the impression is similarly encouraged by the fact that Plato's discussion of the contrast between non-monadic and monadic comprising entities in *Theaetetus* makes no mention of relations between the relevant comprised entities.

Note, however, that an entity's being atomic entails its being monadic. On the other hand, its being monadic does not entail its being atomic.

---

[36] I use the term without intending to connote Leibniz's notion of a 'monad'. Note also that use of 'monadic' to describe comprising entities is intended to have nothing to do with use of the term to describe a type of relation (a type which contrasts with that of polyadic relations).

[37] See Doepke 1984, 46-7; Simons 1987, 212.

[38] See Lewis 1991, 81-87. Note that Lewis uses 'fusion' for the classical mereological notion which I am calling 'sum'.

6. Singleness

There is a fundamental respect in which an entity might be said to be one, which is not captured by the accounts provided for any of the features so far mentioned.

Suppose we wish to consider, for example, the question whether a certain sum of pebbles is identical to the pebbles of which it is the sum.[39] We may imagine that the pebbles are widely dispersed on some beach. Let us assume that the number of pebbles is greater than 1. Thus if the sum of pebbles is identical to the pebbles, it is non-monadic. So I shall take an equivalent formulation of the question to be, whether the sum is non-monadic.[40]

Consider ways in which the sum of pebbles is said to be one, apart from the respect under investigation, i.e. that of monadicity. Let us assume that the sum is not one, in the sense that it is not unified - the relations between the pebbles are not relations of integrity. Furthermore, since the sum comprises at least two distinct pebbles, it is also not one, in the sense that it is not atomic. Clearly, however, the sum is one in the sense that it is linguistically one. For it is obviously correct to say that the sum is one sum.

Now it would seem, intuitively, that the meanings of the expressions 'the sum of pebbles' and 'the pebbles' are distinct. So that if it turned out that the sum of pebbles is non-monadic, and so identical to the pebbles, we should explain the relation between the two expressions (in respect of their semantic properties) as analogous to the relation between 'the morning star' and 'the evening star'. Otherwise the answer to the question whether the sum of the pebbles is identical to the pebbles would be obviously, and uninformatively, affirmative. But it isn't. To assume that it is *obviously* affirmative would be to fail to make sense, for example, of David Lewis's searching discussion of the question in *Parts of Classes*.[41]

Moreover, it seems that one aspect of this difference of meaning between 'the sum of pebbles' and 'the pebbles' is the following: It is part of the meaning of 'the sum of pebbles' that what it is used to refer to is *one*, in a way that it is *not* part of the meaning of 'the pebbles' that what it

---

[39] I introduce the issue of singleness in terms of the example of a comprising entity designated by 'sum'. Precisely the same problems, however, arise in connection with other types of comprising entity, e.g. collections, wholes, etc.

[40] The questions are equivalent assuming that if the sum is not identical to the pebbles then *a fortiori* it is not identical to any other entities which are greater than 1 in number.

[41] See Lewis 1991, 81-7.

is used to refer to is one.[42] However, the notion of being one required here could not be that of atomicity (for the sum comprises more than one pebble). Nor could it be that of relational oneness (for the sum of pebbles is said to be one, not the pebbles). Similarly, it could not be the notion of unity (for we have assumed that the sum of pebbles, in the particular case we are considering, is not unified). Nor, again, could it be the notion of monadicity (for it is precisely the question whether the sum of pebbles is monadic which is under investigation; to assume that it is in setting out the problem would be to beg the question).

Of the features so far explicated, in association with claiming that an entity is one, only linguistic oneness remains as a candidate for helping to account for the difference in meaning between 'the sum of pebbles' and 'the pebbles'. It seems, however, that this notion too is not able to bear the weight of the task. Consider, for comparison, the contrast between 'the orchestra' and 'the players'. The expression 'the orchestra' is syntactically indeterminate, so that in some contexts it combines with verbs in the *plural* to form grammatical sentences (while combining in other contexts with verbs in the *singular* to form grammatical sentences). It seems that in such contexts the meanings of the two expressions do *not* differ as regards being one. In a context in which we say that the orchestra *are* playing, the meaning of 'the orchestra' does not seem to connote a oneness which 'the players' does not. So the difference we are looking for cannot be explained by reference to linguistic oneness. Something can be linguistically one without displaying the oneness we are inquiring about.

It appears then that the fact that the sum of pebbles is linguistically one cannot explain the difference in meaning between 'the sum of pebbles' and 'the pebbles', as regards being one. Therefore, we are led to conclude that a consideration of the question whether the sum of pebbles is monadic presupposes that the sum of pebbles is one, in a sense distinct from all of those discussed so far.

I shall describe an entity that is one in this other sense as being 'single'. The question that faces us is how to account for this notion of a single entity, which I have so far characterised only negatively (i.e., by saying that being single is not the same as being linguistically one, or atomic, etc.). The account I wish to propose is a rough and tentative one,

---

[42] Lewis makes such an assumption when he urges (as a reason for finally denying that a sum is identical, strictly speaking, to the parts which compose it) that "what is true of the many is not exactly what's true of the one. After all they are many while it is one" (Lewis 1991, 87).

and yet I hope it suffices to show that the notion of a single entity is not hopelessly obscure. In any case, I do not pretend to have shown conclusively that we are forced to acknowledge some such notion. The larger argument of the present work will not depend on the assumption that such a notion must be acknowledged, nor, in particular, that the account I offer for this notion is an adequate one.

It should be noted, however, that if the notion singleness is rejected, then when we compare, as regards being one, what is designated by different expressions for comprising entities - for example 'plurality of the organism's cells' and 'organism' - we can only contrast them with regard to one of the features so far accounted for. So that if, for example, both are linguistically one, both are non-atomic, both are equally unified (the organism and the plurality of cells comprise the same cells, and since the cells are integrally related to one another, any entity which comprises them is unified), and both are monadic (as I have noted above, on some views any comprising entity must be deemed monadic), then there would seem to be no sense in which the organism is one, according to which the plurality of cells is not one.

I propose, then, to account for the notion of a single entity as follows. Tokens of referring expressions, such as 'the horse', or 'those houses', can be characterised as bearing a *semantic relation* to what they are used to refer to. I shall assume that such semantic relations always belong to one of two types: They are either one-one relations, or one-many relations. That is, with respect to the positions they occupy in the structure of such a semantic relation, the referring expression is always one, and what is being referred to is either one or many.

I am assuming, in particular, that neither the referring expression, nor what is being referred to, can be *indeterminate* in this respect: Neither of them can be either indeterminately one or indeterminately many. This assumption seems to be very natural, and is principally supported by the obscurity which would be involved in the alternative of allowing an indeterminateness in this respect.

Expressions (again, in the sense of *token* expressions) associated with a one-one semantic relation I shall describe as 'one-one expressions', whereas those associated with a one-many relation I shall describe as 'one-many expressions'. Paradigmatic examples of one-one expressions may be found among expressions involving terms designating atomic entities of certain types, in the singular, such as 'the electron'. As regards one-many expressions, corresponding paradigmatic examples would be expressions involving terms designating atomic entities in the *plural*, such as 'the electrons'.

The notion of a single entity can now be explained, on the basis of my assumptions regarding the semantics of referring expressions. I propose that an entity is single if and only if it can be referred to by means of a one-one expression. Contrastingly, an entity will be described as 'multiple' if and only if it can be referred to by means of a one-many expression.

Thus suppose a token of the expression 'the sum of pebbles' can be used to refer to a certain sum of pebbles. If this expression is one-one, then the sum of pebbles is a single entity. If the expression is one-many, then the sum of pebbles is a multiple entity. In the former case, the relation between the expression 'the sum of pebbles' and the many pebbles is *mediated* by the comprising entity, the sum of pebbles. The expression is *semantically related* to the sum of pebbles, and the sum of pebbles *comprises* the pebbles. In the latter case, the expression 'the sum of pebbles' is semantically related directly to the many pebbles. The contrast between the cases is illustrated in Figures 1a and 1b below. Semantic relations are represented by dotted arrows, and comprising relations by continuous arrows.

Case 1: 'the sum of pebbles' is one-one
(implying that the sum of pebbles is a single entity):

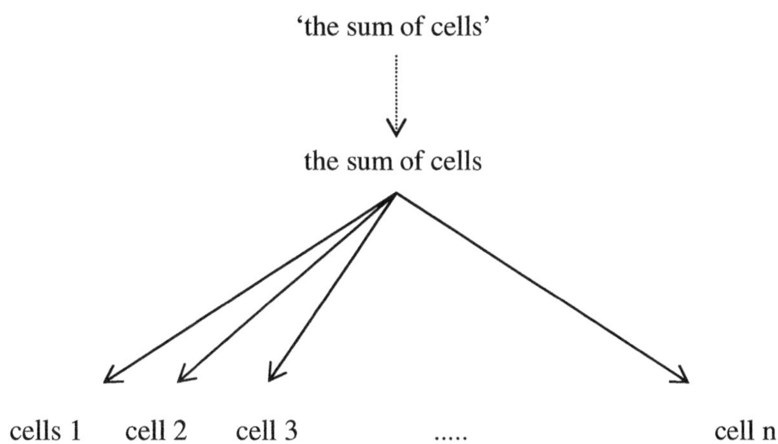

Figure 1a

Case 2: 'the sum of pebbles' is one-many
(implying that the sum of pebbles is a multiple entity):

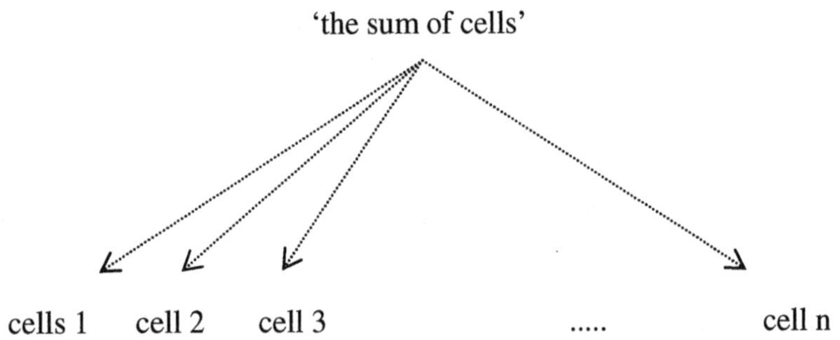

Figure 1b

Note that I have not assumed that being single is incompatible with being multiple. In Case 1, it follows from our assumptions that the sum of pebbles is a single entity. It does not follow, however, that the sum of pebbles is not a multiple entity. For it to be a multiple entity, all that is required is that *some* one-many expression can be used to refer to it. It is not required that 'the sum of pebbles' be that expression.

If being single is not incompatible with being multiple, then we may consistently assume that the sum of pebbles is a single entity and the pebbles are a multiple entity, and yet the sum of pebbles is identical to the pebbles. Thus the expressions 'the sum of pebbles' and 'the pebbles' may have a different meaning (as regards being one), and yet they may be used to refer to the same entity. In this way the possibility of an informative, non-obvious answer to the question whether the sum of pebbles is identical to the pebbles is accounted for.

Having explained the notion of singleness in terms of the distinction between expressions that are one-one and expressions that are one-many, the question now arises whether a criterion is available for distinguishing between expressions that are one-one and expressions that are one-many. Consideration of our example case of 'the sum of pebbles' and 'the pebbles' suggests a simple and attractive proposal.

We noted that what is referred to by means of the expression 'the sum of pebbles' is thought of as being one (in a way that, ultimately, we accounted for in terms of the notion of singleness). What is referred to by means of the expression 'the pebbles', by contrast, is correspondingly thought of as being many. It seems clear, on reflection, that this difference in the way we think about the referents of these expressions corresponds to a difference in the way we predicate things of them. Thinking of the referent of 'the sum of pebbles' as being one, in the relevant way, corresponds to saying that the sum *is* on the beach; thinking of referent of 'the pebbles' as being many corresponds to saying that the pebbles *are* on the beach.

The proposal, then, amounts to this: a token expression is one-one if and only if it is used (in the context of a grammatical sentence) in combination with a (certain kind of) verb, or more generally, verb phrase, in the singular. It is one-many, on the other hand, if and only if it is used (in the context of a grammatical sentence) in combination with a (certain kind of) verb, or verb phrase, in the plural.

The unspecific qualification 'certain kind of' is meant to exclude a true statement such as 'the sum is a sum' from being taken to indicate that 'the sum' is here one-one, and so as evidence that the sum is a single entity. The qualification can be partly fleshed out by saying that any case in which the use of the expression in combination with a verb in the singular is guaranteed merely by the *linguistic oneness* of the entity referred to should not count as evidence for the referring expression's being one-one. One might admit that the orchestra *is* an orchestra, even in a context in which one insists on speaking of the orchestra using verbs such as 'play' in the plural. In such a context, it seems to me that no evidence is provided for the orchestra's being a single entity.

If the relevant distinction between kinds of verbs (or verb phrases) cannot ultimately be justified, then the qualification 'certain kind of' will have to be dropped, and complying with the criterion for linguistic oneness will guarantee complying with the criterion for singleness.

In what follows, I shall assume the correctness of this criterion without further argument. A critical examination of the suggestion would take us far afield into questions concerning the relations between the syntactic and semantic structure of a language - an investigation which lies well beyond the scope of the present work. The fact that this assumption is not strongly supported will be of little consequence, since, as noted before, my account of singleness does not play a crucial role in my larger argument. Its role is chiefly that of offering an additional perspective which might help to grasp the significance of the claims I make later. Ultimately, if the account of singleness I offer is rejected, then the additional perspective is lost, but the soundness of the general argument should not be affected.

This concludes my account of different respects in which a comprising entity might be said to be one. Armed with these distinctions regarding ways of being one, we are in a much better position to assess claims which contrast different types of comprising entity by saying that an entity of one type is 'more one' than an entity of the other type. For we may examine each of the entities, feature by feature, and see in which respect each of the entities is one. If they do not differ either as regards linguistic oneness, or atomicity, or unity, or monadicity, or singleness, then we shall conclude that initial impressions notwithstanding, they do not differ with respect to being one.

I arrive at precisely such a conclusion, for example, in comparing collections with sums. It might be assumed at first glance that a sum, as conceived of in classical mereology, is 'more one' than a corresponding collection (i.e. a collection whose members are identical to entities of which the sum is a sum). However, according to a plausible account of the notion of a collection, which I outline later in this chapter, a collection is one in every respect in which the corresponding sum is one.

I shall be comparing three general conceptions of wholes, in the course of Parts Two and Three of this work - as classical mereological sums, as neoclassical mereological sums, and as Unities. The availability of clear accounts as regards respects in which an entity can be said to be one helps us to clarify the relations between these conceptions. It turns out that monadicity, rather than unity, singleness, atomicity or linguistic oneness, is a crucial feature in connection with such a clarification. I shall argue that sums, as conceived of in classical mereology, are non-monadic entities. Many of the advantages of what I call the neoclassical conception of wholes, over the classical one, have to do with precisely those features of neoclassical sums which render them monadic. However, these are at the same time features which are highly problematic. Thus, I shall cast doubt on the view that neoclassical mereology offers a satisfactory conception of monadic wholes. The theory of Unities, which I introduce in Part Three, offers an alternative conception of a monadic comprising entity, which is free from many of the difficulties afflicting the neoclassical conception.

Appendix: Plural Quantification

A study of comprising entities has at the centre of its concerns a relation between many comprised entities and one entity which comprises them. While I have assumed the possibility, in principle, that the comprised entities will be identical to the comprising entity, one cannot sensibly investigate the relation between comprised and comprising entities without carefully distinguishing between talk about the former and talk about the latter. In some contexts one may alternate between the use of phrases like 'the pebbles are nicely arranged' and 'the pebble collection is nicely arranged' without engendering misunderstandings. In a context in which what one is concerned with is precisely the relation between the pebbles and the pebble collection, it is obvious that one cannot do so.

Moreover, a study of comprising entities requires attention not only to things said about *each* one of the comprised entities, but also to things said about *all* of them, claims which cannot be directly rephrased as claims about each of the comprised entities. The claim that those four people weigh a thousand pounds is an example of such a claim, if it is understood in the sense according to which the weight of the people *taken together* is a thousand pounds.

Such claims are most naturally expressed by means of sentences which contain *plural terms*, i.e. referring terms which combine (only) with verbs in the plural, such as 'Sarah and Suzy' or 'the man's bodily organs'. Singular terms, by contrast, are terms which combine (only) with verbs in the singular. Standard predicate calculus makes it possible to express formalised sentences which contain singular terms but not plural terms. To facilitate the expression of formalised sentences which contain plural terms, the standard calculus may be supplemented with the means for plural quantification, that is, with plural variables and quantifiers. [43]

Thus in addition to
singular variables: x, y, z, u, v, w

[43] My exposition here is indebted to, and follows rather closely, that of van Inwagen 1990, Section 2.

We have
plural variables: the *xs*, the *ys*, the *zs*, etc.

The expressions 'the *xs*', 'the *ys*', etc. are not treated as composite expressions, any more than the expressions for singular variables are. For our purposes it is best to think of them as plural pronouns, used in the same way as 'they' or 'them'. It will be convenient to allow the plural variables to be written in an alternative way in some contexts: 'those *xs*' instead of 'the *xs*', etc. Note that 'the *xs*' is not taken to be constructed using '*x*'. The fact that one of the *marks* used in writing the former is the mark used in writing the latter is to be treated as a typographical coincidence.

Using a singular variable, we might say '*x* is in the house'. Similarly, using a plural variable, we might say 'the *xs* are in the house'. More generally, 'the *xs* are *F*' (where 'are *F*' is a predicate). The expression 'is one of' is used in its natural sense, and so I will take it to be clear how sentences of the form '*x* is one of the *ys*' are to be understood. In terms of the relation *is one of*, the following useful types of sentence can be defined:

each of the *ys* is $F =_{\text{def.}}$
for all *z*, if *z* is one of the *ys*, then *z* is *F*

the *xs* are among the *ys* $=_{\text{def.}}$
for all *z*, if *z* is one of the *xs* then *z* is one of the *ys*

the *xs* are properly among the *ys* $=_{\text{def.}}$
the *xs* are among the *ys*, and for some *z*, *z* is one of the *ys* but not one of the *xs*

the *xs* are (severally) identical to the *ys* $=_{\text{def.}}$
for all *z*, *z* is one of the *xs* if and only if *z* is one of the *ys*

An important ambiguity must be noted in connection with sentences of the form 'the *xs* are *F*'. According to one reading, which I call the 'distributive' reading, 'the *xs* are *F*' is equivalent to 'each of the *xs* is *F*' (see the first of the definitions above for an explanation of 'each of '). According to the other reading, which I call the 'collective' reading, 'the *xs* are *F*' is not equivalent to 'each of the *xs* are *F*'.

For example, if the sentence 'the four people weigh a thousand pounds' is read distributively, it is taken to mean that each of the four people weighs a thousand pounds. If, on the other had, it is read collectively, it is taken to mean that the (arithmetical) sum of the four people's weights is equal to a thousand pounds. Note that claims such as 'the *x*s make up *y*' and 'the *x*s compose *y*' are generally read in a *collective* sense.

Turning to quantifiers, we have
singular existential quantifiers : 'for some *x*', 'for some *y*', etc.

and
singular universal quantifiers: 'for all *x*', 'for all *y*', etc.

In addition, we have
plural existential quantifiers: 'for some *x*s', 'for some *y*s', etc.

and
plural universal quantifiers: 'for any *x*s', 'for any *y*s, etc.

<u>Examples</u>:

The sentence

(1) someone weighs a thousand pounds

is formalised as

(1') for some *x*, *x* is a person and *x* weighs a thousand pounds

The sentence

(2) some people weigh a thousand pounds

is ambiguous, however.

It might be formalised as

(2') for some *x*s, each of those *x*s is a person and each of those *x*s weighs a thousand pounds

or as

(2'') for some *x*s, each of those *x*s is a person and those *x*s weigh a thousand pounds (i.e., taken together)

The sentence

(3) all entities have sums

is again ambiguous.
It is either to be read as

(3') for all *x*, for some *y*, *y* is a sum of *x*

or as

(3'') for any *x*s, for some *y*, *y* is a sum of the *x*s

Finally, we should note the different types of predicates which occur in contexts of plural reference. Singular reference makes use of singular *n*-adic predicates, that is, *n*-adic predicates which contain *n* free singular variables. Such predicates express *n*-ary relations. Thus 'is larger than' is a dyadic predicate, containing two free singular variables, which expresses a corresponding relation whereby one thing is larger than another thing. Plural reference makes use of plural predicates in addition to singular predicates. These plural predicates are known as *variably polyadic predicates*, containing free plural variables. The relations they express are known as *multigrade relations*. These names are meant to indicate that such relations, in many cases, hold between variable numbers of entities. Thus the variably polyadic predicate 'are carrying', which features in the sentence 'some people are carrying a piano', expresses a relation between a variable number of entities (it applies, for instance, to cases where two people are carrying a piano, to cases where three people are carrying a piano, etc.).

Chapter 3
Types of Comprising Entities

I - Collective Classes and Distributive Classes

An important step in the development of a theory of what I have called 'controversial' comprising entities (entities designated by terms such as 'class', 'manifold', 'collection', 'aggregate', etc.) is the clear distinction between talk of the many comprised entities, on the one hand, and talk of the entity which comprises them on the other hand. It seems natural to consider the latter as simply a peculiar mode of the former, one in which the many are 'spoken of as one'. Indeed, it may turn out that the comprising entity is identical to the many comprised entities. However, the question whether this is so can only be considered with clarity if our thought and talk of the former is clearly distinguished from our thought and talk of the latter.

In the 19th Century gradually more attention was being paid to comprising entities, due to the emerging importance of their role in connection with the logical analysis of language and with work on the foundations of mathematics. This led to a recognition (for example, in the works of Cantor, Schröder and Frege) of the need for a distinction between talk of a comprising entity and talk of the many entities it comprises.

Once it is clear that talk of a comprising entity cannot simply be assumed to be a peculiar kind of talk of the many comprised entities, it is also clear that existence of the comprising entity cannot without justification be supposed to be merely an aspect of the existence of the many comprised entities. It is then realized that, granted the existence of the many comprised entities, an *additional* assumption is required to guarantee the existence of the comprising entity. The general assumption playing this role in Cantorian set-theory, for example, is the Comprehension Principle, according to which, as we have seen in section IV of Chapter 1, whenever there exist entities *x*s, there exists a class to which the *x*s belong as members.[1]

[1] See also Machover 1996, 12.

In addition to laying down assumptions regarding conditions under which a comprising entity exists, the development of a theory of comprising entities involves the introduction of assumptions regarding the relations between comprised entities and entities which comprise them.

A particularly noteworthy development in this respect was the distinction, due principally to Frege, between two different types of comprising entity, corresponding to two different types of relations between comprised and comprising entities. Frege criticised Schröder for failing to distinguish between two notions of a class:[2] The first is the notion of a class as what Frege calls a 'collective whole'. In specifying such a notion of classes, we 'do not bring them into connection with concepts'.[3] To see what is meant by 'bringing them into connection with concepts' it is perhaps best to contrast this notion of a class with the other one discussed by Frege.

The second notion is that of a class as the *extension* of a concept. The extension of the concept *man* is an entity which is internally articulated in accordance with the concept, in a way that the collective whole is not. A collective whole is an unarticulated occupant of a specified spatial domain. The difference between the two notions is highlighted by the fact that there are two ways in which an entity can belong to a class-as-extension: It can belong to it as a member, or as a subclass.[4] By contrast, there is only one way in which an entity can belong to a class-as-collective-whole: It can only belong to it as a part.

The relation *is a subclass of* is transitive, as is the relation *is a part of*. The relation *is a member of*, by contrast, is not transitive. Furthermore, both *is a subclass of* and *is a part of* may be assumed to be reflexive,[5] but not so the relation *is a member of*. In as far as a theory of classes is meant to serve as a tool in the formalisation of logic, it is classes-as-extensions that are required, for it is only such classes that allow us to avoid conflating, for example, the logical form of 'Socrates is a human-being' with that of 'the Greeks are human beings'.

---

[2] See Frege 1895.

[3] Frege 1895, 87.

[4] As Frege notes (Frege 1895, 92), Husserl similarly observes that there are two ways in which an entity can belong to a class-as-extension, in his 1891 review of the same work of Schröder's work.

[5] To take them as reflexive relations is to assume that an "improper part" is a part, and that an "improper subclass" is a subclass.

Frege's distinction can be thought of as marking a point of bifurcation, at which modern theorizing on classes became separate from modern theorizing on wholes and parts (in a sense which involves the provision of a systematic theory of the relation of parthood).

His classes-as-collective-wholes are predecessors of contemporary classical and neoclassical mereological sums, and his classes-as-extensions are predecessors of contemporary classes and sets.Classes-as-extensions were soon to meet grave difficulties centring on Russell's well-known Paradox, which led to considerations limiting the application of the Comprehension Principle. One way (adopted by Russell) to effect such limitations was by means of introducing a theory of types. Another (adopted by Zermelo and his followers) was to construct an *axiomatic* set theory, where only a limited range of instances of the Comprehension Principle were assumed.[6]

On the other hand, reflection on these difficulties led Stanisław Leśniewski to develop a theory which did away with classes-as-extensions, or *distributive* classes, as he called them, and only made use of Frege's other type of class, which he called *collective* classes.[7] He named his theory of collective classes *mereology*, the etymology of the term expressing the theory's interpretation as a theory of parts and wholes (as already noted, '*meros*' is the Greek for 'part'). In what follows, I shall adopt Leśniewski's terminology, speaking of Frege's distinction as the distinction between distributive and collective classes.

From the perspective of the common origin of modern theories of classes (i.e. theories of distributive classes) and of parts and wholes (i.e. theories of collective classes), it is understandable that Leśniewski did not consider mereology to be any more ontologically presumptuous than the set or class theories of Cantor, Frege, and Russell and Whitehead. Where the latter theories assume the existence of an entity (i.e. a class) which has certain individuals as members, mereology assumes the existence of an entity (i.e. a mereological sum) which has those individuals as parts. That is, both theories assume the existence of a single comprising entity, and they only differ with respect to the formal characteristics of the relations between the comprising entity and the individuals it comprises.

Moreover, the fact that class theory could only be safeguarded against contradiction by apparently placing limitations to the Comprehension

---

[6] See Machover 1996, 13-14.
[7] See Küng 1963, 105-115; Simons 1987, 101-104.

Principle which have often been viewed as ad hoc, whereas mereology could be proven to be consistent without recourse to such limitations,[8] seemed to show that mereological comprising entities were if anything more acceptable than class-theoretical ones. It might be surprising, therefore, that the existence of mereological sums has generally been treated with more suspicion than the existence of sets or classes.

The explanation for this suspicion seems to lie in the fact that the distinction between distributive and collective classes came to be generally perceived as a distinction between abstract and concrete entities, respectively.[9] Our identificational intuitions seem to be led into conflict when we consider (alleged) arbitrarily specified concrete comprising entities, in a way that they do not seem to be with regard to corresponding abstract comprising entities.

If we assume for example that there exists a concrete entity which comprises the car here and a hat on the other side of town, we feel unclear about the circumstances under which we should judge that this alleged comprising entity ceases to exist, in a way that we do not either with regard to the car or with regard to the hat. If 90% of the hat has been destroyed by fire, it might be clear to us that the hat has ceased to exist. Has the alleged comprising entity ceased to exist as well? Classical mereology maintains that a sum of the *x*s exists if and only if all of the *x*s exist. According to this principle, the sum of the car and the hat has ceased to exist. Our intuitions, however, have very little to say either for or against this claim, and it would seem no less plausible to claim that a concrete entity which comprised the car and the hat has continued to exist. Since our identificational intuitions are on unfirm grounds with regard to such alleged comprising entities, it is natural to doubt their existence.

The claim that mereology is no less justified in assuming with respect to any individuals *x*s that an entity exists which has them as parts than set-theory is in assuming that an entity exists which has those individuals as members is emphasized by Nelson Goodman in 'A World of Individuals'.[10] He explains that he is using 'individual' in a technical sense. Various associations arguably conveyed by the term in its ordinary

---

[8] See Simons 1987, 110 (n. 23).

[9] We shall presently find that there are good reasons for doubting that distributive classes need necessarily be assumed to be abstract. However, it is likely that concrete distributive classes would be met with suspicions similar to those associated with concrete collective classes.

[10] Goodman 1956.

use, such as, for instance, that of indivisibility, are explicitly excluded from Goodman's usage. Using 'individual' in this technical sense is, he says, as unobjectionable as using 'class' in the technical sense required in mathematical and logical contexts. In such contexts it is usually supposed that there exists a class whose only members are, say, the planet Mars, the tree next to me, and Henry's right hand, even though common discourse would tend to reject the existence of a class in such a case. Similarly, according to Goodman, (classical) mereology assumes with equal justification that there exists an individual which is the sum of the same individuals (i.e. the planet, the tree and the hand).

It should be noted, however, that the analogy between mereology and set-theory is imperfect. It may be admitted that 'individual' is not less justifiably used as a technical term than 'class' is. However, set-theory does not presuppose that the entities comprised by a class are themselves classes, whereas mereology does presuppose that the entities comprised by an individual are themselves individuals. Thus mereology carries the implication that the comprising entity belongs to the same ontological category as the comprised entities, an implication which is surely of more than merely terminological significance.

II - Distributive Classes and Concreteness

I have said that contemporary conceptions of wholes can be viewed as descendants of Frege's classes-as-collective-wholes, and so as versions of the notion of a collective class. This might incline one to assume dogmatically that concrete wholes (a notion which I will be looking at in detail in the next chapter) are necessarily to be conceived as collective classes. Since the central proposal of the present work is to conceive of concrete wholes in a way which renders them *distinct* from any version of a collective class, it is important for my purposes to undermine any such dogmatic inclination.

With this in mind, it will be instructive to see that it is far from clear that concrete wholes cannot be conceived of as *distributive* classes, rather than as collective classes. Not that I will be urging the conception of concrete wholes as distributive classes. The notion of a Unity which I introduce in Part Three is neither a version of the notion of a collective class nor of the notion of a distributive class. However, by recognizing that it would not be obviously implausible to propose that some concrete wholes are distributive classes rather than collective classes, one becomes accustomed to the idea that although the recent history of the theory of wholes might suggest the contrary, the notion of a concrete whole need not necessarily be understood in terms of the notion of a collective class.

It is not a difference in principle regarding ways in which a distributive class and a collective class can be one, which has led to the view that concrete wholes are collective classes. There seems to be no reason why a class of either type cannot be either atomic or non-atomic, and either unified or ununified (Frege's distinction makes no assumption regarding the number of entities comprised by a class of either type, nor regarding any relations between the comprised entities). Classes of either type are linguistically one (for a class, whether distributive or collective, is one class), and are single entities (for a class of either type is generally spoken of using verbs in the singular).[11] Finally, it is not clear why it should be impossible for a distributive class to be non-monadic, if it is possible for a collective class to be - unless it is assumed that a distributive class (whose members are concrete) must be an abstract entity. I shall return presently to the question as to whether a distributive class must indeed be abstract.

It might be thought that concrete wholes must be conceived as collective classes rather than distributive classes, because the relation of a part to a whole is transitive, which fits *is a part of* (as characterised in connection with the notion of a collective class) but not *is a member of* (as characterised in connection with the notion of a distributive class). However, it may be doubted whether transitivity is indeed an essential feature of the relation *is a part of*. It has been argued, for example, that a certain nucleus is not a part of the heart, even though this nucleus is a part of a certain cell, and the cell is a part of the heart. And similarly, that a handle is not a part of the house, even though the handle is a part of the

[11] In Leśniewski's theory, a distributive class is conceived of as what I have a called a 'multiple' entity.
Leśniewski argues that expressions used in his system which are apparently used to refer to distributive classes should be interpreted as 'shared names', making definite reference to the many class members, rather than to the class. That which is referred to by means of a shared name is what I have called a 'multiple entity'.
Leśniewski, however, took his account to constitute the claim that distributive classes *do not exist*. He clearly equated the view that a distributive class exists with the view that a distributive class is a *single* entity. See Küng 1963, 108-115.

door and the door is a part of the house.[12] If that is so, perhaps such wholes are better understood as distributive classes, allowing the relation between the cell and the heart to be considered as *non*-transitive.

Examining the ways in which classes of the respective types may be one, and the question whether the relation *is a part of* is transitive or not, does not seem to provide us with compelling reasons for thinking that a concrete whole must be viewed as a collective class and not as a distributive class. A more forceful consideration, however, is associated with the *concreteness* of concrete wholes. For there seem to be reasons to think that while collective classes may, in principle, be concrete, distributive classes may not. If that is so, then obviously a concrete whole cannot be a distributive class.

A terminological clarification is called for, before discussing this issue. I confine the term 'class' (on its own, i.e. not in combination with 'distributive' or 'collective') to entities studied in any familiar type of set-theory.[13] In as much as the set-theoretical classes share the essential features of distributive classes, I shall consider them as distributive classes of particular types. Any of the familiar types of set-theory, however, specifies features of sets which are not among those which I noted above as distinguishing a distributive class from a collective class (for example, the assumption that a null class exists). Consequently, although I take it that all classes are distributive classes, I do not assume that all distributive classes are classes.

Now, classes are generally considered to be abstract entities. Since such entities are what most naturally springs to mind as falling under the concept of a distributive class, it is natural to suppose that abstractness is an essential feature of distributive classes. As for collective classes (in cases where such entities comprise concrete individuals), it is clear that from the outset they were assumed by theorists to be concrete entities.[14] Let us take this traditional view regarding collective classes for granted.

---

[12] The argument regarding the example of the heart is found in Rescher 1955; that regarding the example of the house is found in Cruse 1979. Doubts regarding the transitivity of *is a part of* can also be found in Lowe 1989, 95 (n. 9).

[13] Note that in some types of set theory the term 'set' is used to refer to all such classes. In set theories of other types a distinction is drawn, among entities designated by 'class' in this wide sense, between entities designated by 'set' and entities designated by 'class' in a narrower sense. See, e.g. Machover 1996, 16.

[14] For example, Frege assumes that a wood is a collective class of many trees; see Frege 1895, 87. For Leśniewski's similar view, see Küng 1963, 105f.

Let us question, however, whether indeed distributive classes are essentially abstract.

There seem to be two main reasons for thinking that classes are abstract.[15] One is that it is assumed in set-theory that there exists a null class, a class, that is, which has no members. The other is that it is assumed in set-theory that a singleton (i.e. a class which has one and only one member) is not identical to its only member.

It may seem that if a class has no members, then it must be conceived either as an abstract entity, or perhaps not even as an abstract entity but rather, simply, as a fictitious entity.[16] As for the second point, it also seems to lead to the view that classes are abstract, because if the singleton of a pebble is an entity distinct from the pebble, there seems to be no concrete entity distinct from the pebble which might be a plausible candidate for the role of the pebble's singleton.

Without further argument these two points are certainly not compelling. If it is possible that there should exist concrete atomic entities,[17] then it is not clear why a class which has no members should necessarily be abstract.[18] And if it is plausible to claim that a statue is different from the lump of clay which constitutes it, and yet to claim that both the lump and the statue are concrete entities, it may seem plausible to attempt to conceive of the singleton of a pebble as a concrete entity which occupies the same place at the same time as the pebble, without being identical to the pebble. However, let us grant, for the sake of argument, that the two points can be developed convincingly, or that other satisfactory arguments can be produced to show that classes are indeed abstract entities.[19]

What is important to see is that even if we grant that abstractness is an essential feature of classes (that is, of set-theoretical entities), it by no means follows that it is an essential feature of other types of entity falling

---

[15] See, for example, Russell 1919, 183; Grossmann 1983, 209-10, 214-216.

[16] Grossmann, for example, claims that the null class is a convenient fiction, and that this was also the position of the originator of set-theory, Cantor, himself. See Grossmann 1983, 214-15.

[17] For a discussion of the existence of such entities, see Simons 1987, 41-43.

[18] Indeed, David Lewis proposes to conceive of the null class as concrete, by identifying it with the mereological sum of all individuals, which is itself a concrete individual. See Lewis 1991, 10-15.

[19] Grossmann provides two further reasons for thinking that classes are abstract (see Grossmann 1983, 209-16). I do not think his reasons are convincing, but there is no need to take issue with him on these points in the present context.

under the general notion of a distributive class. The arguments for the abstractness of classes, which we have noted above, depend on the assumptions made in set-theory, that null classes and singletons exist. These assumptions, however, depend on the acceptance of specific basic assumptions (expressed as the Comprehension Principle or various restricted versions of it) regarding types of conditions under which it may be assumed that classes exist. If these assumptions were appropriately altered, the existence of the null class and of singletons would not be guaranteed.[20]

To illustrate that distributive classes might be assumed to exist which are not susceptible to the arguments leading us to think that classes are abstract, let us consider a 'reduced' set-theory as follows. According to this theory, the relations *is a member of* and *is a sub-class of* have the same formal properties which they do in familiar set theory. However, the existence of only a very limited range of classes is admitted in the reduced theory, as implied by the following assumptions (which replace the more familiar unrestricted or restricted versions of the Comprehension Principle):

A1 for all *y*, *y* is a class only if *y* is not an individual

A2 for all *y*, *y* is a class if and only if for some *x*s,
y is a class whose members are all and only the *x*s.

A3 for any *x*s, for some *y*, *y* is a class whose members are all and only the *x*s if and only if the *x*s meet the following conditions:
(1) each of the *x*s are individuals
and
(2) for some *u* and for some *v*, *u* is one of the *x*s and *v* is one of the *x*s and $u \neq v$

[20] In the case of Zermelo-Fraenkel set-theory, for example, the existence of a *class* with a single member is guaranteed by the assumption that for any 'definite' property *P* there exists a class $A = \{x: Px\}$. The existence, regarding any individual or set *a*, of a *set* whose only member is *a*, is guaranteed by the latter assumption together with the Axiom of Pairing and several definitions. These assumptions also guarantee the existence of an null class and of an null set. If we accepted only the existence of classes corresponding to *some* 'definite' properties *P*, those consequences would not follow. See Machover 1996, 15-18.

According to this theory, if we assumed that there were three pebbles, *a*, *b* and *c*, and that these were the only individuals in the world, then there would be the following four classes (and only these): {*a*, *b*}, {*a*, *c*}, {*b*, *c*}, and {*a*, *b*, *c*}.

The assumptions rule out the existence of a null class (A2) and of singletons (A3). Furthermore, the conjunction of A1 and A3 rules out such classes as {*a*, {*a*, *b*, *c*}} (the latter would need to be viewed as abstract by someone who rejected the possibility that two concrete entities might be in the same place at the same time). Therefore, the main considerations for thinking that classes are abstract do not apply with regard to such a reduced theory of distributive classes.

I do not mean to suggest, of course, that this reduced set-theory is an attractive alternative to more familiar versions set-theory, in as much as we seek in set-theory an adequate foundation for mathematics. What I am suggesting, rather, is that set-theoretical classes are not the only type of distributive class that we might consider, and that other types might not implausibly be regarded as concrete entities.

Evidence that distributive classes have not always been conceived as abstract entities can be found in Bertrand Russell's discussion of classes and aggregates in *The Principles of Mathematics*. For his account seems to imply that what he calls 'aggregates' are concrete distributive classes.

In the chapter 'Whole and Part', Russell distinguishes between two kinds of whole: 'aggregates' and 'unities'.[21] We need only consider aggregates at present. An aggregate is a class 'regarded as the whole composed of all the terms in those cases where there is such a whole'.[22] In other words, the aggregate is what Russell describes as a 'class as one'. In his account of the notion of a class, he argues that a distinction must be drawn between the notion of a class as many and the notion of a class as one. When we speak of the soldiers (sc. all the soldiers belonging to a certain army), for example, we speak of the class of soldiers as many. When we speak, on the other hand, of the army, we speak of the class of soldiers as one.[23] The existence of a class as one is not guaranteed in every case in which there exists a class as many.[24] However, in cases where the

[21] Russell 1903, 137ff.

[22] Russell 1903, 139.

[23] See Russell 1903, 68-9.

[24] See Russell 1903, 68-9. Conditions under which the existence of a class as one must be denied are conditions which lead to contradictions such as the one known as Russell's Paradox. They are discussed in the chapter 'The Contradiction', Russell 1903, 101ff.

class as one exists, it may be regarded as a 'whole composed of all the terms', a whole of the sort he calls an 'aggregate'.[25]

That an aggregate of concrete individuals is itself concrete is suggested by Russell's claim that a class as one is an entity of the same (logical) type as its members. He explains:

> A class as one, we shall say, is an object of the same *type* as its terms; i.e. any propositional function $\phi(x)$ which is significant when one of the terms is substituted for x is also significant when the class as one is substituted.[26]

Thus, for example, if 'x moves' is a propositional function which is significant when the value of x is one of the members of the class (e.g. one of the soldiers), then it is also significant when the value of x is the class as one (i.e. the army). If it is assumed (plausibly) that whenever something neither moves nor is at rest, 'x moves' is not significant when that thing is the value of x, it follows that the army either moves or is at rest. On these assumptions, the army, and in general any class as one (and thus any Russellian aggregate) whose members are concrete entities, is itself a concrete entity.

That an aggregate of concrete individuals is regarded by Russell as a distributive class can be seen from the way he emphasizes that aggregates involve two distinct relations, each of which can be loosely described as a relation between part and whole. These relations correspond precisely to the relations *is a member of* and *is a subclass of*, and therefore Russell's aggregates comply with Frege's criterion for distributive classes. For Russell contrasts the relation that 'holds only between the aggregate and the single terms ... composing the aggregate', on the one hand, with 'the relation to our aggregate of aggregates containing some but not all the terms of our aggregate' on the other hand: For example, the relation of the Greek nation to the human race is different from that of Socrates to the human race;[27] Russell is assuming that the human race is an aggregate, and that entities that it contains may have either of two different types of relations to it: The relation which Socrates has to it (i.e. *is a member of*), and the relation which another aggregate, the Greek nation, has to it (i.e. *is*

[25] Russell 1903, 138-9.
[26] Russell 1903, 104-5.
[27] Russell 1903, 139.

a subclass of). It transpires, therefore, that Russell's aggregates are Fregean distributive classes, and yet that they are nonetheless concrete entities.

III - Collections

In the present section I wish to bring out a notion which underlies a certain familiar, non-technical use of the term 'collection'. On this basis I introduce a particular technical use of the term, which will be of use in later chapters, since it will be found that wholes according to one conception are collections in this sense, while according to another conception they are not. And so this notion of a collection will provide a perspective from which to illuminate contrasts between different conceptions of wholes. In addition, a version of this notion of a collection will feature in the theory of Unities presented in Part Three, alongside the notion of a Unity itself.

It is of interest to discuss collections at this early stage, before we begin to discuss the notions of part and whole in detail, because the notion of a collection seems to be the source of a certain constraint which has been accepted - mistakenly in my view - as constitutive of the notion of a whole. The reason why this constraint seems to have been accepted, I suggest, is that wholes have tended to be conceived as collections, in a broad sense which I explain below. In as much as to conceive of wholes in this way involves an explication of the notion of a whole, at least in some respect, *in terms of* the notion of a collection, it seems appropriate to discuss collections prior to a discussion of wholes. In one important respect, the present work can be taken as an exploration of the possibility of elaborating a conception wholes according to which this constraint is not accepted.

Collections, I assume, are entities which can be referred to by means of expressions of the form 'the collection ... ', such as 'the collection of pebbles', 'the collection comprising these paintings', etc. I assume that if *y* is a collection, then for some *x*s, *y* comprises the *x*s, so as to make it appropriate to say that the *x*s *form y*. I shall take the claim that *y* is a collection of the *x*s to be equivalent to the claim that *y* is a collection formed by the *x*s. Each of the *x*s, in such a case, is said to be a *member* of *y*.[28]

I assume further that in some contexts it might be said that a collection of meteorites is drifting in space, and is about to collide with a planet; or that a colourful collection of pebbles is found on the beach. Collections spoken of in this way are taken to be entities that are linguistically one, single, and concrete. For 'collection' is used here in combination with a singular indefinite article '*a*' (indicating linguistic oneness), with verbs in the singular, such as 'is' (indicating singleness); and collections are said to have such properties, or relations to other things, as *drifting*, *colliding with*, *being found in* (indicating concreteness).

I conclude that the notion associated with one of the familiar uses of the term 'collection' is the notion of a linguistically one, single and concrete entity (concrete, that is, if the entities it comprises are concrete).[29] To say this much about the notion is, I take it, uncontroversial. By contrast, it is far less easy to say what constraints (if any) follow from this notion of a collection as regards atomicity, unity or monadicity, and also as regards being a distributive class or a collective class. Instead of investigating the issue of these further features at this point, however, I propose to take the notion of a collection as indeterminate in these respects, and to consider further determinations of the notion as corresponding to particular *species* of collections.

Thus there is, for instance, a species of collections which are essentially non-atomic. And there is a species of which being ununified, being non-monadic, and being a distributive class are essential features. At this stage we are investigating the generic notion, and we need not be concerned with the question whether collections belonging to any particularly determined species of this notion actually, or even possibly, exist. At present, therefore, in speaking of a collection I am speaking of a concrete comprising entity, which is single and linguistically one, an entity which may or may not be atomic, unified or monadic.[30] A certain

---

[28] It is not assumed, however, that the relation designated by 'is a member of' in connection with collections is the same as the relation designated by the same term in connection with classes.

[29] This is not to deny that there are other uses of 'collection' according to which a collection may lack one or more of these features.

[30] Notwithstanding the indeterminacy of the notion in these respects, according to the account I have offered so far, it is already sufficiently determinate to be contrasted with the notion Russell took 'collection' to connote in Russell 1903, 69.By assuming as he does that a collection is referred to by means of an expression of the form 'A_1 and A_2 and ... and

additional constraint does seem to follow, however, from the way 'collection' is used in the contexts I have in mind.

Suppose at a time *t* we find pebbles *a*, *b*, *c*, *d* and *e* at some distance from one another on an otherwise pebble-less beach. Suppose we assume that *a*, *b*, *c* and *d* form (at *t*) a collection, which we call C_1. Suppose we also assume that *a*, *b*, *c*, *d*, and *e* form (at *t*) a collection, which we call C_2. I submit that it is true on purely conceptual grounds, given these assumptions, that $C_1 \neq C_2$. If someone agrees with us in assuming that collections C_1 and C_2 are identified by such descriptions, and yet claims that it is possible for C_1 to be identical to C_2, we should say that that person has failed to grasp the concept of a collection and the associated concept of forming,[31] which we have applied in this case.

The qualifying remark 'at *t*' is inserted in brackets, to allow for the possibility that a collection should gain or lose members in time. Ignoring this possibility for a moment, the point just made regarding collections formed by the pebbles may be generalized by saying that collections comply with a principle of Weak Extensionality, defined as follows:

Weak Extensionality

WE for all *u*, for all *v*, for any *xs*, for any *ys*
if (1) *u* comprises the *xs*, and *v* comprises the *ys*
and (2) each of the *xs* and each of the *ys* is an *F*
(where '*F*' is a sortal term)
then
u is identical to *v* only if the *xs* are (severally) identical to the *ys*

This principle rules out the possibility that a five pebble collection should be identical to a four pebble collection (assuming that it is impossible for a collection to gain or lose members). It does not, however, rule out the possibility that a five pebble collection should be identical, for example, to the collection formed by all the molecules comprised by the

---

A_n', Russell fails to guarantee (as I do by means of my assumptions) that a collection is linguistically one, and that it is a single entity (in the sense explained Chapter 2, section II). I grant, however, that this assumption of Russell's is not *incompatible* with the assumption that a collection is single. Furthermore, in as far as he assumes that a collection is a class as one, he would seem to be assuming that it is a distributive class, rather than a collective class. For as I have explained, he seems to assume that a class as one is a distributive class.

[31] I shall assume below that grasp of the relevant concept of a collection involves grasp of the relevant concept of forming.

five pebbles. A principle which would rule this latter possibility out could be arrived at by dropping condition (2). I call this stronger principle - one which I do not assume collections to comply with - Strong Extensionality:[32]

Strong Extensionality

SE for all *u*, for all *v*, for any *xs*, for any *ys*
 if *u* comprises the *xs*, and *v* comprises the *ys*
 then
 u is identical to *v* only if the *xs* are (severally) identical to the *ys*

Principles WE and SE, as stated above, are meant to hold (when they do) irrespective of the time at which *u* comprises the *xs* or *v* comprises the *ys*. WE would clearly fail for collections if it were assumed for example that a collection of four pebbles can acquire a fifth pebble as a member. To accommodate this assumption, we require a temporally relativised version of WE, involving a temporally modified notion of the comprising relation (i.e. the notion of a relation designated by a temporally modified predicate 'comprises at *t*'):[33]

Weak Extensionality - relative to time

WET for all *u*, for all *v*, for any *xs*, for any *ys*, for all *t*,
 if (1) *u* comprises the *xs* at *t*, and *v* comprises the *ys* at *t*
 and (2) each of the *xs* and each of the *ys* is an *F*
 (where '*F*' is a sortal term)
 then
 u is identical to *v* only if the *xs* are (severally) identical to the *ys*

Similarly, a temporally relativised version of SE can be stated as follows:

---

[32] Strong Extensionality is similar to the familiar Principle of Extensionality, generally assumed to hold with regard to *classes* (see Chapter 1, section IV). According to Extensionality, if *u* is a class of the *xs*, and *v* is a class of the *ys*, then *u* is identical to *v* if and only if the *xs* are (severally) identical to the *ys*. Strong Extensionality differs from Extensionality only by replacing the biconditional 'if and only if' with the conditional 'only if'.

[33] I understand the relation between temporally unmodified notions and temporally modified ones in the way explained by Simons (1987, 176-8).

Strong Extensionality - relative to time

SET for all *u*, for all *v*, for any *xs*, for any *ys*, for all *t*,
if *u* comprises the *xs* at *t*, and *v* comprises the *ys* at *t*,
then
u is identical to *v* only if the *xs* are (severally) identical to the *ys*

For the sake of simplicity I shall conduct the discussion in terms of the temporally unmodified notions. It will be a straightforward matter to adapt the conclusions to temporally modified terms. I am claiming, then, that WE is a feature of the concept of a collection, in that the acceptance of WE (for collections) follows necessarily from a grasp of the concept of a collection. The stronger principle, SE, is not in this sense a feature of the concept of a collection.

It seems clear, however, that a principle stronger than WE (although weaker than SE) can be formulated which, again, is a feature of the concept of a collection. Consider again the collection formed by pebbles *a*, *b*, *c* and *d*, i.e. C_1. Now consider a collection formed by *a*, *b*, *c*, *d and* a certain molecule *m* drifting in the air in the vicinity of those pebbles.[34] Call this collection 'C_3'. It would seem that admitting the distinctness of C_1 and C_3 follows from a grasp of the notion of a collection no less than admitting the distinctness of C_1 and C_2 does.

To formulate a principle from which the distinctness of C_1 and C_3 follows, I need to explain first what I mean by saying that the *xs* are *coextensive* with the *ys* (where the *xs* and the *ys* are all entities). The term 'coextensive' is chosen because one of the implications of the *xs* being coextensive with the *ys* is that they collectively - i.e. together - extend over the same region of space as the *ys*. Noting that for *x* to overlap *y* is for *x* to share a part with *y*, I define as follows:

D1 the *xs* are coextensive with the *ys* $=_{\text{def.}}$
for all *z*, *z* overlaps one of the *xs* if and only if *z* overlaps one of the *ys*[35]

---

[34] Perhaps it would be easier to visualise these entities as forming a collection if it is imagined that they are all drifting in proximity to one another in a region of outer space which is otherwise completely empty.

[35] Generally in what follows, the expression 'one of the *xs*' may be assumed to paraphrase 'at least one of the *xs*' unless it is absolutely clear from the context that 'exactly one of the *xs*' is meant.

It can now be proposed that according to the stronger principle we were looking for, which governs the identity of collections, if *u* is a collection of the *x*s and *v* is a collection of the *y*s then *u* is identical to *v* only if the *x*s are coextensive with the *y*s. I am proposing, therefore, that a principle which may be called Coextensive Determination is a feature of the concept of a collection:

<u>Coextensive Determination</u>
CD for all *u*, for all *v*, for any *x*s, for any *y*s
if *u* comprises the *x*s and *v* comprises the *y*s
then
u is identical to *v* only if the *x*s are coextensive with the *y*s

The distinctness of collections C_1 (formed by four pebbles) and C_3 (formed by four pebbles and a molecule) clearly follows from CD. Molecule *m* overlaps one of the members of C_3 (indeed, it *is* one the members of C_3), but overlaps none of the members of C_1. Therefore, C_1 and C_3 are not coextensive, and so according to CD they cannot be identical to one another.
Again, a temporally relativised version of the principle can be formulated. In addition to a temporally modified notion of the relation *comprises*, we require a temporally modified notion of the relation *are coextensive with*. This, in turn, requires a temporally modified notion of the relations *overlaps*, which can be defined in terms of a temporally modified notion of the relation *is a part of*. Thus we may say that *x* overlaps *y* at *t* if and only if *x* and *y* share a part at *t*. And the *x*s are coextensive with the *y*s at *t* if and only if for all *z*, *z* overlaps one of the *x*s at *t* if and only if *z* overlaps one of the *y*s at *t*. We then have:

<u>Coextensive Determination - relative to time</u>
CDT for all *u*, for all *v*, for any *x*s, for any *y*s, for all *t*,
if *u* comprises the *x*s at *t* and *v* comprises the *y*s at *t*
then
u is identical to *v* only if the *x*s are coextensive with the *y*s at *t*

I shall be assuming, then, that collections comply with principle CD (in case their membership is fixed over time) or with CDT (in case their membership varies over time). If a type of comprising entities (e.g. collections) is governed either by CD or by CDT then I shall say that an entity of this type has the feature of being 'coextensively determined'. Collections, therefore, are coextensively determined.

To summarise, collections may be characterised as entities which are linguistically one, single, concrete, and coextensively determined. Both concrete distributive classes, and concrete collective classes, it seems, must be considered to be collections in this sense. For, in as much as distributive classes are generally assumed to comply with Strong Extensionality, and compliance with Strong Extensionality implies compliance with Coextensive Determination (as can be readily seen on inspection), we conclude that *concrete* distributive classes belong to a type of collections. As for concrete collective classes, we shall see - at least to the extent that they are fairly interpreted in terms of the notion of a mereological sum - that they comply with either CD or CDT, and so they too count as collections.

Chapter 4
Theory and Pre-theory of Wholes

I - The Pre-theoretical Conception

In discussing the issue of wholes and parts, we must distinguish between pre-theoretical and theoretical conceptions. Consider, for comparison, a similar distinction that might be drawn in discussing the nature of heat. We have, on the one hand, a pre-theoretical understanding of the notion of heat, an understanding which determines the appropriate use of the term 'hot'. On the other hand, we have theoretical proposals which are intended to explain what the property *is hot* consists in, such as, for example, the (no longer plausible) proposal that it consists in the presence of a type of material substance known as 'caloric', or the alternative proposal that it consists in the molecules making up the hot thing being in such a state as to possess a mean kinetic energy of higher degree than some specified standard. Some of the main tasks of a study of the nature of heat are the introduction and development of, and decision between, such alternative proposals.

The case is similar when we discuss the nature of wholes and parts and the relations between them. Theoretical proposals are put forward with the intention of illuminating a subject matter which is identified by our pre-theoretical understanding of the terms 'whole' and 'part'. The field of such theoretical study is called 'mereology'.

The analogy between the theory of heat and mereology applies notwithstanding the fact that the former is an empirical theory and the latter is a metaphysical theory. In both cases it is a crucial question to what extent the theory is successful in providing an explanatory account of our pre-theoretical notions. It is important to keep this in mind, because one sometimes encounters an attitude towards mereology as if it has made precise, once and for all, our conceptions of whole and part.[1] This is misleading in two ways. First of all, it fails to give sufficient due to the variety of mereological theories that have been proposed, and to the difficulties involved in deciding between them. Secondly, it fails to note

[1] For example, in Lewis 1986a.

that although the range of theories that have been proposed is wide, it is still demonstrably not exhaustive.

Pre-theoretically we speak of a great variety of entities as wholes or as parts. Indeed, there is a broad sense of the term 'whole' according to which it seems to have an almost universal application. For example, we may claim to be considering the building as a whole, or the whole building. And then it seems acceptable to say that the building *is* a whole, although such a claim is admittedly more typical of philosophical (but not only properly mereological) discourse than of common discourse. Similarly the city, the earth, or the universe, can be considered as wholes. Furthermore, it seems to be equally acceptable to speak of the whole tennis match, the whole symphony, the whole proposition, the whole story, or the whole theory. Turning to 'part', we find that it has a similarly wide application - again, if it is taken in a sufficiently broad sense. The building, we may say, is a part of the city, the earth is a part of the universe, the proposition is a part of the theory, the movement is a part of the symphony, and the symphony is a part of the oeuvre of the composer.

With respect to these very general senses of 'whole' and 'part', the chief restriction to their application seems to derive from the way these terms are often taken as *correlative* to one another. To say of x that it is a part is somewhat like saying of u that she is a lover: In both cases a relational attribute is involved. Just as the latter implies that u is a lover *of something*, so the former implies that x is a part *of something*. Generally, the claim that x is a part tends to be understood as implying that there is some y of which x is a part. Correlatively, this y is considered as a whole with respect to x, i.e. a whole one of whose parts is x. Similarly, saying that y is a whole is understood as implying that there is some x which is a part of y, and this x can be described as a part with respect to which y is a whole.

The correlativity of 'whole' and 'part' leads to a restriction of the application of these terms, if we make the further assumption that an entity x cannot be a part of itself - that is, that the relation *is a part of* is irreflexive. In common discourse it seems that 'part' is often understood in such a way as to imply that *is a part of* is indeed irreflexive.[2] If *is a part of* is irreflexive, then saying that x is a part implies that for some y, y is not identical to x, and y is a whole with respect to x. This would suggest, for

[2] Frege seems to take the assumption (by Schröder) that *is a part of* is reflexive to be an unobjectionable *extension* of a more familiar, irreflexive notion. See Frege 1895, 87.

instance, that 'part' does not apply to the universe (if indeed there is an entity which is appropriately referred to by means of the expression 'the universe'), assuming that no entity which is distinct from the universe could plausibly be held to have the universe as a part. Similarly, saying that y is a whole implies that for some x, x is not identical to y, and x is a part of y. On this assumption it would seem that 'whole' does not apply to a geometrical point, for example. For there seems to be no entity distinct from the point which could plausibly be taken to be a part of the point.

In theoretical contexts it is commonly assumed that *is a part of* (by contrast with *is a proper part of*, of which we say more in what follows) is actually a reflexive relation. According to this assumption, the application of 'whole' and 'part' is even less restricted.

We note, then, in the context of the pre-theoretical use of both 'whole' and 'part' a broad sense which is associated with a very wide-ranging application. It is noteworthy that in this sense even classes can be considered as wholes - and this not only with respect to their sub-classes but also with respect to their members. As Rolf Eberle says:

> One speaks of parts and wholes with respect to many kinds of relations. For example, a concrete individual may be regarded as the whole of its qualitative parts, of its physical molecules, or of the events in its history. A set $\{A, B\}$ is a whole composed of the parts $\{A\}$ and $\{B\}$ relative to inclusion, but a whole of the parts A and B with respect to the ancestral of membership.[3]

Indeed, Cantor's famous definition of 'set' has sets described as wholes:

> By a set we are to understand any collection into a whole M of definite and distinguishable objects of our intuition or our thought. These objects are called the elements of M.[4]

Similarly, Russell considered the term 'whole' as appropriate for describing the class-as-one, taking either its members or its subclasses as parts, emphasizing however that the notion of a whole associated with the former assumption is distinct from that associated with the latter

[3] Eberle 1970, 32.
[4] See Burton 1985, 588.

assumption.[5] It may be concluded that the terms 'whole' and 'part', taken in their widest sense, can be used to describe any comprising entity and the entities it comprises, respectively.

In addition to the broad pre-theoretical use of 'whole' and 'part' we have been discussing, there is a narrower pre-theoretical one which takes *concrete* entities as its paradigmatic cases. A stone, a tree, a church, a horse, are all concrete entities, and are all described as wholes (or parts) in this narrower sense. The correlativity of 'whole' and 'part' is a feature of this use as it is of the broader use described above. Thus to say of a tree that it is a whole implies that there is some (concrete entity) x which is a part of the tree - for example, one of its branches. And to say of a stone that it is a part implies that there is some (concrete entity) x of which the stone is a part - say the castle's wall. An abstract entity might be said to be a whole in this narrower sense only if the relations between it and the entities it comprises are taken to be the same as the relations assumed to hold between a concrete whole and its parts.[6]

It is with this narrower pre-theoretical sense of 'whole' and 'part' that the discussion of wholes and parts is usually assumed to be concerned. This explains why the theory of classes is not generally thought to be a part of mereology:[7] The theory of classes is not thought to offer a theoretical account of concrete entities, and in particular not of wholes in the narrower sense. Again, in terms introduced in the former chapter, 'whole' and 'part' in the narrower sense may be generally used to describe a concrete comprising entity and the entities it comprises, respectively. In what follows I shall be assuming, in speaking about wholes and parts, this narrower pre-theoretical usage.

Since different theoretical conceptions of wholes are going to be considered, each associated with its own terminology, it will be useful to make a few comments with a view to clarifying and disambiguating our talk about wholes and parts prior to commitment to one or another theoretical conception.

---

[5] See Russell 1903, 138-9.

[6] Thus Lewis (1991, e.g. 3ff.) argues that classes are wholes in the narrower sense (in particular, that they are mereological sums) with respect to the singletons of their members. This claim must be distinguished from the weaker and much less controversial claim of Eberle's, noted above (in connection with Eberle 1970, 32), that a class can be considered to be a whole, in the *broader* sense of the term, with respect to its members.

[7] Even Lewis, who thinks that mereology *applies* to classes, takes set-theory to involve a non-mereological relation, namely the relation between an individual and its singleton. See, for example, Lewis 1991, 87.

1. 'part' and 'whole'

One may begin by assuming that some entities (in what follows, unless explicitly stated otherwise, I use 'entity' to speak of *concrete* entities) stand in the relation *is a part of* to other entities. If *x* is a part of *y*, then we will say that *y* is a *whole* with respect to *x*. Thus 'is a whole with respect to' may be taken to designate the converse of the relation *is a part of*. The claim that *y* is a whole with respect to *x* is alternatively expressed by saying that *y has x* as a part. And the claim that for some *x*, *y* is a whole with respect to *x*, can be expressed by saying simply that *y* has parts.

Note that all of this is consistent with saying that *y* is a whole with respect to many entities (which are not identical to one another), and with saying that *x* is a part of many entities (which are not identical to one another).

Suppose we have five letter-shaped pieces of plastic, P, L, A, T and O (they have been named after the letter types of which they are tokens, respectively). These plastic pieces may be spoken of obliquely as 'letters'. Suppose these letters are arranged so that a token of the name 'Plato' exists, and let this token be referred to, simply, as 'Plato'. In this case, P stands in the relation *is a part of* to PLATO, so that P is a part of PLATO. Correlatively, PLATO stands to P in the relation *is a whole with respect to*, so that PLATO is a whole with respect to P (or PLATO has P as a part). Indeed, PLATO has parts.

2. 'overlap' and 'disjoint'

If two entities have a part in common then they are said to *overlap* one another. That is, if *z* is a part of *x*, and *z* is a part of *y*, then *x* and *y* overlap one another (this could also be expressed by saying that *x* overlaps *y*). If entities do not overlap one another then we say they are *disjoint* from one another.

Suppose PLA and ATO are entities which are parts of PLATO, where PLA has P, L and A, but neither T nor O, as parts; and where ATO has A, T and O, but neither P nor L as parts. Thus PLA overlaps ATO, and it similarly overlaps P, or L. P, however, is disjoint from ATO, and similarly it is disjoint from L, or O.

3. 'make up'

We say that the letters P, L, A, T and O *make up* PLATO. In pre-theoretical discourse, 'compose' is synonymous with 'make up', and I have used 'compose' in preference to 'make up' so far as the pre-theoretical term (for instance, in Chapter 1). However, since 'compose' is often used in presentations of the *theory* of wholes and parts, that is, in classical and neoclassical mereological contexts, and since one of the main points I wish to make in this study is that this term as used in those contexts is understood in an overly limited sense, I have deemed it better, to avoid confusion, to use the term 'compose' in this familiar technical sense in what follows and to use 'make up' as the pre-theoretical term.

The predicate 'make up' must be understood here as applying collectively, not distributively, to all the letters. The way it applies is therefore analogous to the way 'weigh a thousand pounds' applies, in a given case, to seven people, where it is understood that none of those people weighs a thousand pounds, but that their weights add up to one thousand pounds.

The expression 'make up' (and similarly the pre-theoretical 'compose') suggests things *being put together* so as to produce something, to cause something to exist. It may also suggest things simply *being together* as causally responsible for the existence of something (namely, a whole which has the former as parts). Theoretical reflection on this pre-theoretical term often takes it for granted that if a whole is divided into several disjoint parts (like PLATO), no less (and no more) than *all* of these parts are required to make up the whole. This is no doubt true in many cases. For example, it is quite clear that P, L, A, T and O are all required, no less and no more, to make up PLATO. It will be crucial to my discussion, however, to suspend the assumption that this is always and necessarily the case.

In the theories I shall be calling 'classical mereology' and 'neoclassical mereology', however, this assumption is evidently adopted. It is characteristic of them to interpret the pre-theoretical 'make up' as the theoretical 'compose', defined as follows:

D1 the *x*s compose $y =_{\text{def.}}$
for all *z*, the *z* overlaps one of the *x*s if and only if *z* overlaps *y*

We shall see that this is the relation the *x*s bear to *y* if, and only if, *y* is a sum of the *x*s.[8] In what I shall call the 'theory of Unities', however, the

[8] Van Inwagen is unusual in taking 'the *x*s compose *y*' not to be equivalent to '*y* is a sum of the *x*s'. He takes the *x*s to compose *y* if and only if *y* is a sum of the *x*s *and* the *x*s are all

pre-theoretical 'make up' is not understood in terms of 'compose' in this theoretical sense.

4. 'the parts'

On the basis of the assumptions that have already been explicitly made regarding the case of PLATO and its parts, it is clear that P, L, A, T and O are not the only parts of PLATO. For it has been assumed, for example, that PLA and ATO are parts of PLATO. And of course it is natural to assume also that each of the letters is itself a whole with respect to various chunks of plastic. If P, for example, is broken into two pieces, we are likely to say that each of these pieces was formerly a part of P.

Nevertheless, it is convenient and natural in many contexts to speak of P, L, A, T and O as *the* parts of PLAO. This may be described as a qualified, or context-bound, use of the expression 'the parts'. For often, in considering wholes and parts, one takes the following perspective. One considers that there are many (i.e. two or more) entities, the *x*s, ignoring any parts that they may have. And one considers further that there is an entity, *y*, which the *x*s make up, ignoring any other parts which *y* might have. Since only the *x*s have fallen under consideration so far as parts of *y*, and since they are *all* the parts which have fallen under consideration, it is appropriate in the context to describe them as *the* parts of *y*.

In the discussion that follows it will be assumed that 'the parts' is used to refer to all the entities that are being considered *in that context* as making up the relevant whole. Thus we may freely speak, for example, either of many cells, or of many molecules, as *the* parts of a human organism. Context will determine which of these two options, or indeed of these and many others (e.g. many nuclear particles, or many lumps of flesh and bones, etc.), is the intended one. And if this is not explicitly specified by context, we may still use the expression 'the parts' as referring, indifferently, either to the many molecules, or to the many cells, or to any other plausible candidates for consideration as being comprised by the human organism.

To agree on such a use of 'the parts' is not to rule out, however, the possibility that in some cases 'the parts', or 'all the parts', is used to refer *strictly speaking* to all the parts of the whole under consideration. Used in this way (say, in connection with PLATO), one refers by means of 'the

disjoint from one another. See van Inwagen 1990, 28-29. For a more typical understanding of 'compose', see Lewis 1991, 73-74.

parts' to the *xs*, such that *any* entity that is a part of PLATO is among the *xs* (including, for example, any molecule that is a part of one of the letters). It will be clear from the context which of these different uses is intended.

5. 'proper part' and 'improper part'

A distinction is usually made between two types of parts: Parts that are *proper*, and parts that are *improper*. The pre-theoretical notion of a *proper* part can be ambiguously explained as that of a part which does not include everything included by the whole. An *improper* part is a part which is not a proper part. Thus P, PLA and PLAT are proper parts of PLATO. While if PLATO is at all a part of itself, it must be an improper part. The notions of a proper part and an improper part can made more precise (in a theoretical context), controversially, in terms of the following assumption:[9]

A1 for all *x*, for all *y*, if *x* is a proper part of *y* then
y has a proper part which is disjoint from *x*

Alternatively, it might be proposed that the claim that a proper part 'does not include everything included by the whole' should be rather understood in terms of the following different assumption:

A2 for all *x*, for all *y*, if *x* is a proper part of *y* then
y has a part which is not a part of *x*

Given the assumptions (widely held in mereology) that every entity is a part of itself, but no entity is a proper part of itself, it can be seen that it is consistent with A2, but not with A1, that (for some *x* and some *y*) *x* is the *only* proper part of *y*.[10]

It is clear from the foregoing paragraphs that the notion of an improper part is understood as that of a part which *does* 'include everything included by the whole'. It is usually assumed that this implies that an

---

[9] Simons refers to this assumption regarding the relation *is a proper part of* as the Weak Supplementation Principle. See Simons 1987, 28.

[10] Simons argues that the assumption that an entity might be the *only* proper part of a whole is unacceptable on fundamental, conceptual grounds. I do not find his position on this matter compelling, but I have no reason to take issue with it here. See Simons 1987, 26.

improper part of a whole is necessarily *identical* to the whole. This assumption makes it possible to define 'is a part of' in terms of 'is a proper part of' and 'is identical to', in a way which (as we shall see) is characteristic of classical mereology: [11]

D2 *x* is a part of *y* $=_{\text{def.}}$
x is a proper part of *y* or *x* is identical to *y*

It will be seen, however, that in neoclassical mereology D2 is rejected, together with the assumption that an improper part of a whole is identical to a whole.

6. 'atom'

If a concrete entity has no proper parts, it is said to be an *atom*. I shall assume that atoms not only do not have proper parts, but are such that if any entities, the *x*s, are comprised by an atom *y*, then *each* of the *x*s is identical to *y*. Thus I am assuming that atoms are atomic entities, in the sense of 'atomic' explained in Chapter 2, section I.

Atoms in this sense must not be confused of course with microscopic entities of a certain type which are known by the same name and typically studied in physics. The 'atoms' of physical theory are currently thought to have proper parts (e.g. protons, neutrons, electrons), and to that extent are not atoms in our sense. Note also that an entity's being indivisible (in the sense that it is physically impossible to break it up, to cause parts of it to separate and disperse), does not entail its being an atom in our sense. For it may well be that some individuals have parts which it is physically impossible to separate.

II - Preliminaries to a Theoretical Conception

In Chapter 3 it was noted that two conceptions of a comprising entity emerged towards the end of the 19th Century: That of a distributive class, and that of a collective class. As these conceptions developed, the

[11] See Simons 1987, 11, 26, 37 (D2 is equivalent to Simons' SD1).

assumption established itself that entities of the former type are essentially abstract entities, while entities of the latter type are concrete if the entities they comprise are concrete. Since the pre-theoretical notion of a whole (we speak here of wholes in the narrower sense) is the notion of a concrete comprising entity, it is not surprising that the notion of a collective class became the chief theoretical model according to which wholes were conceived.

It is appropriate to describe the notion of a collective class as offering a theoretical conception of wholes, because it entails claims about wholes and their relations to individuals that are their parts which do not seem to follow incontrovertibly from the rather unspecific pre-theoretical beliefs we have about wholes. For example, if wholes are conceived as collective classes, then the relation *is a part of* is assumed to be a transitive relation, and yet we have already seen in the second section of Chapter 3 that this assumption is controversial.

Indeed, we saw there that there are reasons for thinking that distributive classes are not essentially abstract (although, admittedly, set-theoretical classes are plausibly conceived as abstract entities). If distributive classes of some specified type are concrete entities, then it would seem that the notion of a distributive class offers an alternative theoretical model for wholes. If a whole whose parts are the *x*s is conceivable either as a collective class of the *x*s or as a distributive class of the *x*s, then the notions of a collective class and of a distributive class are alternative theoretical conceptions of the whole, providing alternative accounts of the nature of that whole. Our situation is then somewhat analogous to that of the scientist who needed to decide between two competing theories of heat.

Modern mereology, however, has been by and large a theory of collective classes, commonly known in the context of its discussions as 'sums'. Notwithstanding its varied development and the many forms it has taken, current mereological thinking in general takes core features of the notion of a sum as fundamental. It follows that the received mereological wisdom, in the main, urges the theoretical conception of *wholes as sums*.

According to the conception of wholes as sums, one replaces the pre-theoretical terms 'whole', 'part' and 'make up' with the theoretical, or technical, terms 'sum', 'part' and 'compose', where the assumptions laid down by the theory, using the latter terms, give those terms their specific technical meanings. (Note that 'part' is homonymous; it appears both among the pre-theoretical terms - in a pre-theoretical sense, and among the theoretical terms - in a distinct, theoretical sense).[12]

Turning to the alternative available model for conceiving wholes, namely, the notion of a distributive class, if for convenience we call a concrete distributive class 'Class' (with upper case 'C'), then this alternative can be described as the conception of *wholes as Classes*. According to this conception, one reads for the pre-theoretical 'whole', 'part' and 'make up' the theoretical 'Class', 'member' and 'form', respectively. Again, the meanings of the latter terms are determined by their roles in stating the assumptions of the theory.

This alternative has generally received very little attention. Exceptions to this neglect include, arguably, Russell's theory of wholes in *The Principles of Mathematics*, and Simons' theory of so-called 'concrete classes' in *Parts*.[13] Even if it is granted that both speak of concrete distributive classes, neither of the authors has argued that the conception of wholes as Classes is in any respect preferable to their conception as sums, at least as regards the type of wholes that are generally at the centre of interest of mereological discussions, which I have described as *prima facie* individuals. The case has not been made that entities such as a man, or a table, or a planet, should be conceived as Classes rather than as sums.

The notion of a sum, by contrast, has received much attention and has been widely applied in theorizing about *prima facie* individuals. This wide exposure, together with the apparent absence of an attractive alternative, perhaps explains why it is sometimes forgotten that the conception of wholes as sums, however plausible, is in principle one which might be rejected, or at least one which might in principle be deemed inappropriate in connection with particular types of cases.

Against this background, what I shall be calling the Theory of Unities is to be seen in broad brush-strokes as offering a distinct third model, alongside the two currently available ones of wholes as sums and wholes as Classes. As far as I have been able to ascertain, this third model is completely without modern precedent. It may be described as the conception of *wholes as Unities* (the reason for choosing the term 'Unity' will become clear as we progress).

---

[12] These senses should not be confused with one another. Indeed, in as much as different versions of the conception of wholes as sums involve different assumptions with regard to the formal properties of *is a part of*, they may be said to understand 'part' in different ways from one another.

[13] See Russell 1903, 138-9; Simons 1987, 144ff.

Although the conception of wholes as Classes is not without interest, it will not be treated in the present work. My main interest lies in the conception of wholes as Unities. Most of the discussion that follows is aimed at a comparison between the predominant conception of wholes as sums, and their less familiar conception as Unities. Much of Part Two is devoted to a presentation and discussion of the conception of wholes as sums. Among other points, I shall emphasize the flexibility of this conception, expressed in its ability to accommodate both 'lower grade' notions of wholes, such as those often associated with the terms 'heap' and 'aggregate', and a range of traditional 'higher grade' notions of wholes such as that of a Hegelian organic whole, of a Gestalt-theoretical whole, and of a Husserlian pregnant whole.

Notwithstanding this flexibility, however, I hope to show that the conception of wholes as Unities is not only possessed of an intrinsic interest and plausibility, but also has in certain respects tangible advantages over their conception as sums. Part Three will be devoted, in the main, to this conception of wholes as Unities. According to it, one interprets the pre-theoretical terms 'whole', 'part' and 'make up' (in some contexts, at least) as the theoretical terms 'Unity', 'element' and 'underlie' respectively.

My comments on the different terms used in connection with the three models are summarily presented in the following table:

Model (Wholes as-)	**Comprised Entity**	**Comprising Relation**	**Comprising Entity**
Classes	members	form	Class
sums	parts	compose	sum
Unities	elements	underlie	Unity

The differences between the models are not merely terminological, of course. I shall examine them in some detail in Part Three. It will be seen, for example (to give a quick initial impression of the substantial differences between the models) that in the case of Classes the *x*s may form the same entity as the *y*s only if the *x*s are (severally) identical to the *y*s. In the case of sums, by contrast, the *x*s may compose the same entity as the *ys* only if the *xs* are coextensive with the *ys*. Although Classes and sums differ in this way, they are both *coextensively determined* comprising entities, and so they both count as collections of certain types (see section III of Chapter 3). Whereas in the case of Unities, the *xs* may underlie the same entity as the *ys* even if the *xs* are neither (severally) identical to the *ys* nor coextensive with the *ys*, and indeed, Unities are not coextensively determined comprising entities, and as a consequence, they are not collections.

The distinction between distributive classes and collective classes, which I have said introduces two of the three models which mark our theoretical scenery in outline, has so far been explained in somewhat general terms. In his original analysis of it Frege provides us with additional insights concerning the character of collective classes.[14] Since we will be dealing extensively in the next part of this study with a conception of wholes founded on the idea of a collective class, it is worth our while to give Frege's analysis a closer look.

Suppose the *xs* are pebbles, and *u* is a collective class of the *xs*, and *v* is a distributive class of the pebbles. Let us say that each of the pebbles is a *component* of *u*, and similarly that each of them is a component of *v*. Let us say that a collective class of the *ys*, where the *ys* are among the *xs*, is a *sub-class* of *u*. And similarly, a distributive class of the *ys*, where the *ys* are among the *xs*, is a sub-class of *v*.

Frege pointed out that the two types of comprising entity must be distinguished from one another, because although entities of both types may be said to 'include' both their components and their sub-classes, in distributive classes these two relations of 'inclusion' are distinct, whereas in collective classes they are not. That is, the relation which a component of a collective class bears to the collective class has the same formal characteristics as the relation which a sub-class of a collective class bears to the collective class. Whereas the relation which a component of a distributive class bears to the distributive class has different formal characteristics from those of the relation which a sub-class of a distributive class bears to the distributive class.

It seemed to Frege (for reasons that will be considered presently) appropriate to identify the relation of a component to a collective class as the relation *is a part of*. I use 'is a member of' to express the relation

---

[14] Frege 1895.

Frege takes to hold between a component of a distributive class and the distributive class.[15]

Thus if *u* is a collective class of pebbles, then every component of *u* (i.e. every pebble) and every sub-class of *u* (i.e. every class the components of which are among the pebbles) is a part of *u*. On the other hand, if *u* is a distributive class of pebbles, then every component of *u* is a member of *u*. As regards sub-classes of distributive classes, Frege assumed that the relation each of them bears to the class is precisely the relation *is a part of*.

We have, then, a formal distinction between two relations: the relation *is a part of*, and the relation *is a member of*. Entities that are 'included' in a collective class all bear the former relation to the collective class. As for items that are 'included' in a distributive class, some bear the former relation to the distributive class, others bear the latter relation to it. The formal features by which Frege distinguishes the relation *is a part of* from the relation *is a member of* are expressed by the following two axioms:

A1 Every entity is included in itself

A2 If one entity is included in a second, and this in a third,
then the first is likewise included in the third[16]

Making appropriate substitutions,[17] these axioms apply to the relation *is a part of*, that is, in the case of a collective class, both to the relation between a component and the collective class and to the relation between a sub-class and the collective class. They equally apply in the case of a distributive class to the relation between a sub-class and the distributive class. Importantly, however, the axioms do not apply in the case of a distributive class to the relation between a component and the distributive class.

We have then the following two axioms as (at least partially) characterizing the relation *is a part of* according to Frege:

---

[15] Frege's own term for 'is a member of' in Frege 1895 is 'subter'.

[16] See Frege 1895, 86. Frege uses 'domain' instead of 'entity' at this point of his discussion, adapting to the terminology used in the work of Schröder which he is criticising.

[17] That is, substituting 'is a part of', 'is a member of' or 'is a sub-class of' for 'is included in', as the respective cases require.

A1' Every entity is a part of itself

A2' If one entity is a part of a second, and this of a third, then the first is likewise a part of the third

Axiom A1' expresses the claim that *is a part of* is a reflexive relation, and A2' expresses the claim that it is a transitive relation.

The view that axioms A1' and A2' apply to a relation appropriately described as the relation *is a part of* was apparently suggested to Frege by examples in which the formal characteristics of this relation are evidently realized, examples which it was natural to judge as paradigmatic examples exhibiting the relations between parts and wholes. One such example is given by the areas into which a plane-surface is divided by two (non-parallel) sets of parallel lines. Any two pairs chosen respectively from these sets delimit an area in the shape of a parallelogram. Clearly, some such areas are *contained* (in the spatial sense) in some others. Given our understanding of the notion of spatial containment, it would seem undeniable that axioms A1 and A2 apply (on appropriate substitutions) in the case of this example with respect to the relation *is contained in*, and this relation in turn is naturally described as a relation between a part of an area and an area of which it is a part. Thus it is plausible to take these axioms as capturing a characteristic feature of the relation *is a part of*.

The assumption that axioms A1' and A2' apply to the relation *is a part of* is a natural one, and one that is very widely held. Eberle, for instance, characterizes the most general notion of a relation which is to qualify *prima facie* as a species of *is a part of* by saying that a whole, in the most general sense, may be identified with the supremum of a set relative to an arbitrary partial ordering.[18] A relation is a partial ordering if and only if it is reflexive, antisymmetric and transitive.[19] To say that a whole is the

[18] Eberle 1970, 32. Eberle uses here the term 'sum', but as his foregoing remarks make plain, he takes 'whole' and 'sum' to be synonymous. I use the term 'sum' only for the more specialised theoretical notion characteristic of classical mereology and of theories which follow with regard to certain general principles, as will be explained in what follows.

[19] These features are defined as follows (see Eberle pp. 13-14):

The *field* of a relation R is the set of all items x such that either x bears R to something or something bears R to x. Suppose F is the field of R. Then:

A relation R is *reflexive* if for all x in F, xRx.

A relation R is *irreflexive* if for all x in F, not xRx.

A relation R is *symmetric* if for all x, y in F, if xRy then yRx.

supremum of a set relative to some partial ordering is to imply that the relation *is a part of* is a partial ordering, and thus, in particular, that it is reflexive and transitive.[20] Therefore, according to Eberle, reflexivity and transitivity are features of the relation *is a part of*, taken in its most general sense.

Similarly, Simons states that the principles embodying the claim that the relation *is a proper part of* is a *strict* partial ordering[21] are partly constitutive of 'part', which means that anyone who seriously disagrees with them has failed to understand the word.[22] And again, summarising at the end of his study, he claims that being a strict partial ordering is part of the minimum we can require of a relation if it is to be one of proper part to whole.[23] Simons' subscription to the definitional connections between the relations *is a proper part of* and *is a part of* means that he agrees with Eberle (and with Frege) in taking the relation *is a part of* in its most general sense to be both reflexive and transitive.[24]

We have already noted however that doubts can be, and have been, raised with regard to the assumption that *is a part of* is transitive (and similar doubts are applicable regarding the assumption that *is a proper part of* is transitive). Thus it is not completely clear that A1' and A2' provide us with even *necessary* conditions for a relation to be a relation of the type *is a part of*.

It is clear, however, that even if A1' and A2' correctly characterize such necessary conditions for the relation *is a part of* they do *not* provide sufficient conditions for a relation to be of this type, for reflexivity and

---

A relation R is *asymmetric* if for all x, y in F, if xRy then not yRx.
A relation R is *antisymmetric* if for all x, y in F, if xRy and yRx then $x=y$.
A relation R is *transitive* if for all x, y, z in F, if xRy and yRz then xRz.

[20] Eberle defines 'supremum' as follows:

$\sup_R S$ [the *supremum*, relative to R, of S] = the unique object y which is such that
(1) every member of S bears R to y
(2) for all z, if every member of S bears R to z, then y itself bears R to z
(3) R is a partial ordering
We say that *sup*$_R$ *S exists* just in case there is one and only one object y such that the conditions (1), (2) and (3) are satisfied.

[21] A relation R is a *strict partial ordering* if it is irreflexive, antisymmetric and transitive (see Eberle 1970, 14)

[22] Simons 1987, 11.

[23] Simons 1987, 362.

[24] In fact, Simons considers two notions of *is a part of*, as defined by reference to that of *is a proper part of* (the latter is taken as primitive in his system). See Simons 1987, 26 (SD1) and 112 (SD15).

transitivity characterize many relations that are clearly not species of *is a part of*. For example, the relation *is lighter than or equal in weight to* is reflexive and transitive, but the claim that stone *A* is lighter than or equal in weight to stone *B* is obviously compatible with the claim that stone *A* is *not* a part of stone *B*.

This is not to say, however, that the axioms are not sufficient for Frege's purpose of distinguishing between two distinct relations of inclusion, and establishing that classes, as required for the purposes of the foundations of mathematics, must involve a relation of inclusion that does not fulfil these axioms (namely, the relation *is a member of*). Indeed, the specification of axioms for the relation *is a part of* is only incidental to Frege's, as he was not interested (in the context of the 1895 article) in developing a theoretical account of wholes.

By contrast, classical mereology, versions of which were first introduced by Leśniewski as Mereology and by Leonard and Goodman as the Calculus of Individuals, was explicitly intended by its authors as a theoretical account of wholes. Dissatisfaction with various aspects of set-theory led these authors to develop a theory based on an alternative notion of a comprising entity, that of a collective class.

One of the main reasons why it was reasonable to believe that entities described as collective classes exist was that the characteristics assumed for collective classes seem to accord well with our beliefs about entities of a very familiar type, namely, wholes (in the narrower sense). The interpretation of the notion of a collective class as a theoretically clarified conception of a whole endowed the former notion with a certain plausibility, allowing it to be considered as a serious alternative to the notion of a distributive class, once it was realized that this latter notion was in many ways problematic.

Thus the authors of classical mereology aimed to provide a theoretical account of the notion of a whole, with the expectation that such a theoretical account, conceived as an elaboration of the notion of a collective class, could replace the problematic notion of a distributive class (as developed in set-theory) for a variety of philosophical purposes.

Classical mereology speaks, in a given case, of a comprising entity such that any of the comprised entities is related to the comprising entity by the relation *is a part of*, where this relation is taken to be both reflexive and transitive, as in Frege's account. Thus classical mereology speaks of comprising entities which are understood as collective classes. As we shall see, however, classical mereology has more to say about the way in which such comprising individuals are to be characterized.

Part Two: Sums

Chapter 5
Classical and Neoclassical Mereology

I - Principles of Classical Mereology

Theories of wholes and parts, as I have said, offer particular ways of understanding corresponding pre-theoretical notions. A consideration of comprising entities in general in Part One has suggested that more than one such way may be available, that more than one sort of comprising entity may serve as a model in terms of which to interpret the pre-theoretical notions of whole and part. We now turn to the more detailed discussion of theories which take collective classes as their model. In the context of such theories collective classes are usually called 'sums', and the theories may thus be described as theories which conceive wholes as sums. Every such theory proposes to present systematically the connections between the various pre-theoretical notions associated with wholes, and especially those of being a part or a proper part, of being a whole, and of the parts making up the whole. The latter notion is interpreted as that of the parts *composing* the whole, where the relation *compose* is given a precise analysis in terms of the relation *is a part of* or *is a proper part of*. Thus, in describing wholes and the parts which make up those wholes, these theories speak of sums and the parts which compose them.

Since sums are collective classes, it is to be expected that the assumptions of a theory of wholes as sums should reflect the minimal formal characteristics of collective classes which Frege distinguished, as we saw in the last chapter. Accordingly, a theory of wholes as sums takes the relation *is a part of* to be transitive and reflexive, and takes the relation *is a proper part of* to be transitive and asymmetric (and thus irreflexive). However, as I have pointed out, these minimal formal requirements give us only a partial characterisation of these relations, since other relations which are clearly not relations between part and whole equally comply with these requirements. The relation *is lighter than or equal in weight to*, for example, is also transitive and reflexive. It is clear, therefore, that a systematic theory of wholes as sums will include additional assumptions.

Intuitively, a crucial difference between (1) *a*'s being a proper part of *b* and (2) *a*'s being lighter than *b*, is that (1) involves the idea that *a* is somehow 'in' *b* or 'included in' *b*, which (2) does not. Comprising entities in general are entities which we think of as including the comprised entities in some such way. Now, that *a* is 'in' *b* in the relevant sense but not identical with *b* (recall that *b* is not a proper part of itself), seems to imply something further about *b*, namely, that something *else* is also 'in' it, something in addition to *a*. It seems therefore that we may further characterise the relation *is a proper part of* by the following assumption: If *x* is a proper part of *y*, then *y* has a further proper part *z* which is disjoint from *x* (i.e. *z* shares no parts with *x*). Simons describes assumptions which articulate our intuition that if *y* is not a part of *x* (and thus in particular if *x* is a proper part of *y*), *y* must have a proper part disjoint from *x* - as *supplementation principles*. I shall call the particular assumption stated in the sentence before the last simply the Principle of Supplementation of Proper Parts.[1]

In addition to clarifying the formal characteristics of the relations *is a part of* and *is a proper part of*, theories of wholes as sums are concerned with specifying which entities they take to exist. The general idea is that the notion of a sum, much like the notion of a class, involves a mechanism for 'generating' entities. The Comprehension Principle of Cantorian set-theory, for instance, dictates that whenever there are many entities $x_1, \ldots, x_n$, there is a class which has those entities as its only members. Thus if we begin by assuming the existence of many entities, the notion of a class guarantees the existence of a further entity, and so may be described as involving a principle of generation of entities. The problem with set-theory in the eyes of Leśniewski and Leonard and Goodman, the originators of the first fully developed theories of wholes as sum (that is of classical mereology), was not that it allowed for the generation of entities as such, but rather that it allowed for an overly permissive form of generation. The interest of classical mereology in wholes (conceived as sums) was largely motivated by the fact that wholes seemed to offer a

[1] For a detailed discussion of supplementation principles, see Simons 1987, 26-37. What I have called the Principle of Supplementation of Proper Parts is his Weak Supplementation Principle (p.28).

framework which allows one to generate entities in less excessive ways. Both *is a member of* and *is a part of* may be described as 'generating relations', but the former generates entities in a way that results in an incoherent ontology, whereas the latter does not.

Nelson Goodman explains his difficulty with set-theory by saying, first of all, that it violates the ontological principle of parsimony according to which 'there should be no distinction without distinction in content'.[2] The notion of content relevant here is the notion of the ultimate entities which analysis reveals to be responsible for generating the entity under consideration. Suppose, for example, *a*, *b* and *c* are atoms. Consider classes $A=\{a, b\}$, $B=\{a, b, c\}$, and $C=\{A, c\}$. According to the Principle of Extensionality, classes *A* and *C* are distinct from one another, since they have different members. However, *A* and *B* are entities which Goodman would describe as having the same content. For they are both generated by precisely the same atoms, *a*, *b* and *c*.

Second, the ontological extension that set-theory involves is in his view extravagant and implausible, in that the assumption that an entity exists leads to the assumption that an *infinite* number of entities exist. For if individual *a* exists, then according to set-theory a singleton of *a* exists, and this singleton is not identical to *a*. Furthermore the singleton of *a*'s singleton exists, identical neither to *a* nor to its singleton; and a singleton of the second singleton exists, etc. Note that here too we have a violation of Goodman's principle of parsimony, for each of the infinitely many singletons in this hierarchy has precisely the same content as the others.

That classical mereology does not violate Goodman's principle is evident from the way it defines sums, in contrast with the way set-theory defines classes. Consider for example three individuals *a*, *b* and *c*, and compare the class and the sum which they respectively 'generate'. The class is defined simply as 'the class whose only members are *a*, *b* and *c*'. The sum, by contrast, is not defined as 'the sum whose only parts are *a*, *b*, and *c*'. For if *a* for example has a part *d*, distinct from each of *a*, *b* and *c*, then by the transitivity of *is a part of* it must also be a part of the sum. Describing the sum as 'the sum whose only parts are *a*, *b* and *c*' cannot be correct, therefore. What classical mereology suggests instead is that we define the relation *is a sum of* thus:

---

[2] See Goodman 1956.

D1 y is a sum of the xs $=_{\text{def.}}$
z overlaps y if and only if z overlaps one of the xs

In other words, a sum y of the xs is an individual such that whatever overlaps it overlaps one of the xs, and whatever overlaps one of the xs overlaps it.

This definition allows us to assume that an individual B that is a sum of a, b, and c is identical to an individual C that is a sum of A and c, where A is a sum of a and b (by contrast with what would be the case in connection with classes). For whatever overlaps B overlaps C, and whatever overlaps C overlaps B. Similarly, the definition allows us to assume that an individual D is identical to an individual a, where D is a sum of a. For again, whatever overlaps a overlaps D, and whatever overlaps D overlaps a.

Classical mereologists nevertheless intended their theory to emulate the generative powers of set-theory, for they were hoping to use mereology as a substitute for set-theory in various contexts. Thus, by analogy with set theory, classical mereology assumes that any arbitrarily specified individuals generate one and only one sum. It assumes, in other words, what I shall be calling the Principle of Universal Existence of Sums, which is analogous to Cantorian set-theory's Comprehension Principle, and the Principle of Uniqueness of Sums, analogous to set-theory's Principle of Extensionality. These assumptions, together with the others we have mentioned, are stated explicitly in the list below which we may considered as defining basic principles of classical mereology.

It should be noted that classical mereology is an axiomatic system: The theorems of classical mereology are all derivable from a specified set of axioms and definitions. As is often the case with such systems, different axiom bases, and correspondingly different sets of definitions, might be chosen, all of which are equivalent in being equally sufficient for deriving the same body of theorems which is classical mereology. An example of such differences between equivalent axiom bases can be found in comparing the system presented by Nelson Goodman with that presented by Simons.[3] In Goodman's system, 'overlaps' is taken as primitive, and 'is identical to', 'is a part of' and 'is a proper part of' are defined in terms of it, as well as expressions designating other notions such as 'sum' and 'product'. In the system of Simons, by contrast, 'is a proper part of' and

[3] See Goodman 1951, ch.2; Simons 1987, 25-41.

'is identical to' are taken as primitive, and 'overlaps', 'is a part of', and other expressions are defined in their terms.

In presenting the principles of classical mereology, my main aim is to arrive at a list which makes it possible to contrast classical and neoclassical systems in the most natural and perspicuous way. Simons' list of definitions and axioms provides a very helpful point of departure for developing such a list. However, one of the purposes guiding the construction of his list is that of bringing together all and only those definitions and axioms required to provide a basis from which the theorems of classical mereology can be derived. This is not my purpose here.

First of all, I leave out definitions which introduce notions that we have already introduced as belonging to the pre-theoretical framework for the discussion of wholes and parts. Thus I leave out the definitions which introduce the notions of overlap and disjointness. Secondly, I leave out definitions of mereological terms which will not occur in our discussion, such as 'product', 'difference', 'universe' and 'complement'. Thirdly, I add to Simons' axioms an assumption (P6) which is derivable from his definitions and axioms (and thus one which is not logically independent of the axioms and definitions), but which needs to be added explicitly in some neoclassical theories which do not accept all classical principles. Describing definitions and axioms generically as *principles*, the following list of principles of classical mereology is arrived at:

Definition of 'part' in Terms of 'proper part' and 'identical'

P1 *x* is a part of *y* $=_{\text{def.}}$
x is a proper part of *y* or *x* is identical to *y*

Definition of 'sum'

P2 *y* is a sum of the *xs* $=_{\text{def.}}$
for all *z*, *z* overlaps *y* if and only if *z* overlaps one of the *xs*[4]

Asymmetry of *is a proper part of*

P3 for all *x*, and for all *y*, if *x* is a proper part of *y*
then *y* is not a proper part of *x*

[4] By saying that *z* overlaps *one* of the *xs* it is meant that *z* overlaps *at least* one of the *xs*.

Transitivity of *is a proper part of*
P4 for all *x*, for all *y*, for all *z*,
if *x* is a proper part of *y* and *y* is a proper part of *z*
then *x* is a proper part of *z*

Supplementation of Proper parts
P5 for all *x*, for all *y*, if *x* is a proper part of *y*
then for some *z*, *z* is a proper part of *y* and *z* is disjoint from *x*

Uniqueness of sums
P6 for all *u*, for all *v*, for any *xs*, if *u* is a sum of the *xs* and *v* is a sum of the *xs* then *u* is identical to *v*

Universal Existence of Sums
P7 for any *xs*, for some *y*,
y is a sum of the *xs*

Principles P1 and P2 correspond to Simons' definitions SD1 and SD9, respectively. P3, P4, P5 and P7 correspond to Simons' axioms SA1, SA2, SA3 and SA24, respectively.[5] The principle of uniqueness of sums, P6, is one which Simons takes to follow from his definitions and axioms.[6]

II - Limitations of the Notion of a Classical Sum

Classical mereology is a very 'democratic' theory of wholes and parts, in that it requires for any *xs* that there be one and only one entity which is their sum. There are no particular conditions which need be met by the *xs* in order for there to be a sum of the *xs*. The principle of Universal Existence of Sums (P7) guarantees that a proton tens of thousands of kilometres below the surface of the Sun, and the fly buzzing away here in the room, for example, have a sum.

[5] See Simons 1987, 37.
[6] Lewis describes Principle P6 as one of "the basic axioms of mereology" (see Lewis 1991, 74).

That is, there is an entity *y* of which one part is buzzing here peacefully and another part embroiled in radically different circumstances, very far away, with vast heaving masses and great expanses of open space between them, masses and expanses which are not themselves parts of the same individual. When, in the course of time, the molecules which are now parts of the fly become widely dispersed as parts of the atmosphere, the earth, and many and varied animals and plants, an entity *z* which is the sum of those molecules and the proton far away in the sun will still exist (assuming that all those molecules and the distant proton still exist), and this sum is identical to *y*. The identity of *z* to *y* is guaranteed by the principle of Uniqueness of Sums (P6), given the identity of the entities of which both *y* and *z* are a sum.

1. The Aggregate Argument

Perhaps the most familiar objection to such a conception of wholes is that it is incompatible with the common observation that the existence of at least some types of entity depends on parts of those entities being related to one another in certain ways. This consideration dates back at least as far as Aristotle's Aggregate Argument:[7] In the case of at least some types of entity, namely, entities that are *prima facie* individuals, if the parts of the entity are sufficiently dislocated, separated from one another, so that certain spatial and functional relations between them are disrupted, then these parts may all continue to exist, and yet the entity of which they are parts ceases to exist.

For example, we might imagine a tree all of whose molecules are separated from one other and dispersed over a wide region, as a consequence of certain physical forces coming to bear on the tree (e.g. such as would be engendered by a proximate nuclear explosion). Clearly it is plausible to claim that the tree ceases to exist as a result of this dispersal of its parts. According to the classical mereological view, however, an entity is identified with the unique sum of its parts, and the sum exists if and only if the parts exist, so that the sum exists in cases of the type just mentioned both before and after the dislocation of the parts. It is a consequence of the classical mereological view, then, that an entity exists both before and after the dislocation of its parts, in conflict with the

[7] See Aristotle, *Metaphysics* 1041b11-31. The term 'Aggregate Argument' is used by Scaltsas to refer to this argument. See his discussion in Scaltsas 1994a, 64-5, and Scaltsas 1992, 196-8.

plausible point made by the Aggregate Argument. Let us consider responses which a classical mereologist can make to this point.

2. The Four-dimensionalist Response

A classical mereologist might deny the claim that entities may persist, strictly speaking, through time. Concrete entities must be conceived according to this view as occurrents rather than continuants.[8] If this is assumed, then nothing which exists strictly speaking at some time *t* can exist at some other time *t*'. In particular, the molecules of a tree which exists at *t* cannot be identical to certain widely dispersed molecules which exist at *t*', and so the sum of the former cannot be identical to the sum of the latter, the classical mereological assumptions notwithstanding.

This widely held position is sometimes described as the 'four-dimensionalist view', since according to it trees and other familiar entities are taken not as they seem to be - three-dimensional entities that persist through time - but rather as four-dimensional entities which have time as their fourth extensional dimension.

Four-dimensionalism, as we see, supports a classical mereological conception of wholes, and it is not surprising therefore to find typical classical mereologists such as Goodman and Lewis upholding four-dimensionalism.[9] Adopting four-dimensionalism, however, does not provide the classical mereologist with a solution to a similar difficulty that might be raised in modal rather than temporal terms. Consider an example in which the molecules of a tree *might be* (rather than *are*) dispersed at *t*. If the tree is identical to the sum, then the tree would exist even if the molecules were dispersed (for on classical mereological assumptions, the sum would exist in that circumstance). But clearly, the tree would not exist under those (counterfactual) conditions, and so the tree cannot be identical to the sum.

The appeal to four-dimensionalism does not resolve the difficulty in this case, because what is at issue are not those properties possessed by the tree and the sum at different times, but rather modal properties, i.e. properties possessed by them in different possible worlds. To reply to this modal version of the Aggregate Argument, the classical mereologist

---

[8] The contrast between continuants and occurrents is explained in Chapter 2, section I.

[9] See Goodman 1956; Lewis 1991 (e.g. 77). Note, however, that Whitehead according to Simons is an example of a four-dimensionalist who is not a classical mereologist, in that he denies the assumption of Universal Existence of Sums (P7); see Simons 1987, 81-6.

would have to provide an account which denies that individuals have such modal properties.[10]

More importantly, four-dimensionalism involves a rejection of one of our most deeply rooted intuitions about the world and about ourselves, namely that some entities do not have a merely instantaneous existence, but persist in their existence over a period of time. The deep roots of these intuitions reveal themselves in the difficulties which are encountered when we try to spell out in detail the way in which the world might be described in a *purely* four-dimensionalist language.[11] A physicist who resorts to four-dimensionalism as providing a convenient framework for the interpretation of an abstractly formulated physical theory, may not need to worry about the availability of such a purely four-dimensionalist language. The ultimate untenability of four-dimensionalism would threaten not the *truth* of the physical theory (which is assessed on empirical grounds), but rather only the way in which the theory is interpreted. A metaphysician, by contrast, must be worried about this, for her interest focuses precisely on such empirically un-evaluable issues.

It will be seen in Part Three that an argument involving mereological considerations, which has sometimes been taken as providing a compelling reason to embrace a four-dimensionalist view,[12] can be answered by revising one's mereology in preference to revising one's conviction concerning the persistence of entities through time. In any case, I shall not pursue the question of four-dimensionalism any further in context of the present work. Leaving it aside, I wish only to register my lack of sympathy for four-dimensionalism, and my assumption that it stands in favour of a theory if it allows us to avoid resorting to that view.

3. An Alternative Classical Mereological Response

A different defence of classical mereology, in the face of the Aggregate Argument, can be made without recourse to four-dimensionalism. According to this defence, no entity ceases to exist on the dispersal of the tree's molecules. Rather, there is a property, *is a tree*, which a sum of molecules may acquire or lose. At some point in the history of the sum it

---

[10] This point, with regard to a similar problem, is made by Simons in Simons 1987, 122.

[11] For a discussion of difficulties at making sense of a four-dimensionalist account, see Chisholm 1976, Appendix A; van Inwagen 1987, 36-7 ; Simons 1987, 122-127.

[12] I have in mind Mark Heller's discussion (in Heller 1988) of the puzzle which I treat below in Chapter 10, section III. A variety of views on the same puzzle is surveyed by Simons (1987, 117-121).

turns into a tree, and at some later point it ceases to be a tree. Incidentally, it is likely to be not very much later, given the rapidity with which molecules which make up what we usually take to be one and the same tree are replaced. Suppose S_1 is a sum of the molecules m_1, m_2, ... and m_n, such that S_1 is a tree at t_1. Suppose one of the molecules, m_n, is lost at t_2, perhaps by becoming a part of an insect. Then at t_2 S_1 cannot be a tree (assuming that no part of a tree can be a part of an insect), although S_2, a sum of molecules m_1, m_2, ... and m_{n-1}, may well be a tree at this time.

This view is somewhat similar to the one Frederick Doepke describes as the Reductivist View.[13] The latter view rejects the arguments for distinguishing (for example) between a gold ring and the lump of gold which constitutes it by claiming that strictly speaking, only the ultimate particles making up the ring should be held to exist. Gold rings, and similarly trees, are not real entities, but logical constructions out of real entities.[14]

The defence of classical mereology we are currently considering is not committed to a position quite as strong as the Reductivist View, for this defence acknowledges the existence not only of ultimate particles but also of arbitrarily specified sums of such particles. The classical mereologist can claim that there exists an entity (i.e. a sum of the relevant ultimate particles) which is a tree at t, or an entity which is green at t, claims that are not available to the Reductivist because none of the ultimate particles are either trees or green.

The classical mereologist who argues along the lines suggested, however, is nevertheless radically revisionary of common sense. For the argument requires us to deny that entities fall into the kinds recognised by common sense. The world, according to the view that transpires, does not contain entities which are appropriately described as trees, horses, mountains, buildings or chairs. The sum of such and such ultimate particles might be a tree at t_1, a car at t_2, a horse and a collection of three boulders at t_3 (or any one of indefinitely many other possibilities). Thus the sum is no more an entity of one of these kinds than of another, and

---

[13] See Doepke 1984, 57-60.

[14] Doepke, as well as Simons (both following Wiggins), argue that the ring and the lump are distinct entities which occupy the same region of space at the same time. See Wiggins 1967, Doepke 1984, Simons 1987, 112-7, 210ff. Chisholm's view is similar to the Reductivist View, in that he takes entities which are not susceptible to flux of parts to be *entia successiva*, which he considers to be constructions. See Chisholm 1976, ch.3 (note esp. p. 98); Appendix A.

since only ultimate particles and their sums exist, no entity which can be described as belonging to one of these kinds can be said to exist. The view implies, therefore, that the only existing kinds of entity are kinds of ultimate particles and kinds of sums of ultimate particles.

The question as to whether such a view is ultimately tenable or not, I leave as lying outside the scope of the present inquiry. In any case it seems to me that a mereological theory that would force such a view on us, on purely *mereological* grounds, should not be eagerly accepted. At least not before the possibility of devising a metaphysically less limiting mereological theory is carefully looked into.

4. Problems with Uniqueness of Sums

The Aggregate Argument in the form presented above emphasizes problems associated with the principle of Universal Existence of Sums (P7). A similar argument may be presented, however, which seems to point to problems for the principle of Uniqueness of Sums (P6).

A syllable (see the next paragraph, regarding the type/token ambiguity here), it is said, cannot be identified with the classical sum of the letters which compose it. For example, the syllable SO cannot be identified with the classical sum of S and O. This is not only because according to P7 a sum of the letters exists even in a case where the letters are not appropriately juxtaposed to one another as for example in the case in which the letters are part of the expression 'das Obst' (just as much as the tree does not exist if the molecules are widely dispersed). The syllable SO cannot be identified with the classical sum of the letters for the further reason that those letters compose more than one syllable, i.e. both SO and OS. Since the sum is unique, according to P6, and since SO and OS are different from one another, at most one of the syllables could be identified with the sum. But there would seem to be no more reason for identifying the sum with SO than with OS. Thus we should conclude that neither SO nor OS can be identified with the classical sum of S and O.[15]

Thus formulated, however, the argument is misleadingly vague about whether it speaks of types or tokens of syllables and letters. Leaving aside relations between types as outside the domain of our discussion, it is difficult to present a case along the lines of this argument for the existence of letter-tokens which have more than one sum at some time *t*. As Lewis

[15] The example is taken from (but used to make a slightly different point that made in) John McDowell's discussion of Plato's *Theaetetus*, (1973, 241).

shows in *Parts of Classes*, several possible clarified versions of this argument do not lead to a violation of the principle of Uniqueness of Sums (P6), if one is willing to pay the theoretical price indicated in the foregoing paragraphs in connection with preserving the principle of Universal Existence of Sums (P7), e.g. if one is willing to be committed to four-dimensionalism.[16]

A stronger case for the violation of the principle of Uniqueness of Sums, however, might be made in connection with examples such as that of a clay statue and the lump of clay which constitutes it, or that of a person and the organism which constitutes her/him. A clay statue and the lump of clay which constitutes it are associated with different criteria of identity, respectively. Evidence for this is that, for example, the lump is judged to survive certain deformations which result in the destruction of the statue. Both the statue and the lump, however, would seem to be sums of the same clay particles (for something overlaps either of them if and only if it overlaps one of the clay particles), so it seems that wholes such as statues cannot be adequately accounted for as unique classical sums. Various responses to this argument are available to the classical mereologist, responses, however, which typically rely on controversial accounts of identity and of modality.[17]

5. Problems with Mereological Constancy

We have looked at problems which arise in connection with particular principles among those assumed in classical mereology, P6 and P7. A noteworthy difficulty with classical mereology, however, is associated with a feature of the theory that is not expressed in some particular principle, so much as in the use of *temporally unmodified* predicates in formulating the theory. This implies that the relation *is a proper part of* (as well as other mereological relations defined in its terms) is treated as if entities bear this relation to one another timelessly, in the way entities which are not thought of as being in space and time are taken to bear relations to one another (e.g. in the way the number 8 bears the relation *is greater than* to the number 5).

Typical concrete entities are normally assumed to be capable of losing or gaining parts. It is assumed to be possible, for example, for a molecule which is a part of a tree at t, not to be a part of the tree at t' (where $t \neq t'$).

---

[16] See Lewis 1991, 78-9. See also a related discussion in Simons 1987, 113-4.
[17] See for example Noonan 1993.

The fact that classical mereology makes use of temporally unmodified predicates, however, renders it impossible even to formulate such common sense claims in its terms. The terms of classical mereology are adequate, at best, for expressing claims about entities that are *mereologically constant*, that is, entities which have whatever parts they have as long as they exist. They are unsuitable for expressing claims about entities that are *mereologically varying*.

Indeed it is controversial whether the common sense assumption that entities may lose or gain parts is to be upheld in metaphysical theory. The assumption is clearly incompatible with four-dimensionalism, but it is rejected also by theorists who reject four-dimensionalism. One such theorist, Roderick Chisholm, for example, has argued for a thesis of 'mereological essentialism', according to which the parts of an entity - that is, of a *genuine* entity - are essential to it.[18] This implies, as he says, that 'if *y* is ever part of *x*, *y* will be part of *x* as long as *x* exists'.[19] Genuine entities, therefore, are mereologically constant, and an account of them does not require temporally modified mereological predicates.

Even Chisholm, however, requires *some* notion of the relation *is a part of* which incorporates temporal modification (namely, the notion which applies to entities which are not genuine, i.e. to *entia successiva*).[20] And to the extent that one does not think it appropriate to accept four-dimensionalism or some other reductive view of the sort touched upon above, merely as a consequence of the fact that the terms in which a theory is formulated are unsuitable for describing mereological variation - to that extent one is bound to be critical of classical mereology in this respect.

This brief and selective sketch of problems met with by an adherent of classical mereology should suffice to indicate why its rejection might seem attractive to many authors. I am not concerned in the present framework, however, to settle the issues which weigh for and against the acceptability of classical mereology as an account of wholes and parts, but rather to flesh out of what the classical and neoclassical theories involve and how they contrast with one another. I wish to show that the contrast

[18] Chisholm distinguishes between genuine entities, which are necessarily mereologically constant, and so-called '*entia successiva*', which may well be mereologically varying, but are reducible to genuine entities, and are to viewed as logical construction upon genuine entities. See Chisholm 1976, 98; 154.
[19] Chisholm 1976, 145.
[20] See Chisholm 1976, 155.

between them is compatible with their sharing certain features. These are features which pertain to all theories which account for wholes as sums, whether classical or neoclassical. I shall presently be noting further problems with classical mereology which are associated with the general features of theories of wholes as sums.

III - Flexibility of the Notion of a Neoclassical Sum

It is important to realise that a criticism of classical mereology is not necessarily a criticism of the conception of wholes as sums. The discussion of the last section suggests that some of the most striking difficulties that arise in connection with classical mereology as a theory of wholes and parts are associated with two of the principles which it lays down: The Principle of Uniqueness of Sums (P6), and the principle of Universal Existence of Sums (P7). The notion of a sum as defined above, however, entails neither of these principles. Moreover, neither of these principles is entailed by the conjunction of principles P2-P5. Principles P2-P5 (or their temporally modified analogues - of which more below) may be taken, therefore, as providing the core for what might be described as '*neoclassical*' mereologies. Any mereology which is weaker than the classical one, asserting P2, P3, P4 and P5 (or their temporally modified analogues), but not all of the principles P1-P7, I shall designate accordingly as 'a neoclassical mereology'.

As for the omission of P1 from the neoclassical core, the reason for this can be seen as follows.[21] By denying P6, one opens up the possibility that two distinct entities may be sums of precisely the same parts. For example, it becomes possible for a gold ring to be distinct from a lump of gold, and yet at the same time to be sums of the same molecules. In that case the ring and the lump would be *superposed* with one another, for we define:

[21] I find Simons' explanation of reasons for omitting P1 somewhat unclear. See Simons 1987, 112; 180; and 210-11.

D1 *u* is superposed with $v =_{\text{def.}}$
for some *xs*, *u* is a sum of the *xs* and *v* is a sum of the *xs*

Moreover, it becomes possible for the ring to be distinct from the lump even if both entities have exactly the same proper parts. In that case the ring is not only superposed with the lump, but actually *coincides* with it, given the definition:

D2 *u* coincides with $v =_{\text{def.}}$
for all *x*, *x* is a proper part of *u* if and only if *x* is a proper part of *v*

The need to deny P1 arises, however, even if one is to justify the weaker claim that the ring is superposed with the lump. For it seems correct to say not only that the ring is a sum of the molecules, but also that it is a sum of the lump (for something overlaps the ring if and only if it overlaps the lump). And if the ring is a sum of the lump, then the lump is a part of the ring.[22] But the lump cannot be a *proper* part of the ring, according to the principle of Supplementation of Proper Parts (P5), for the ring has no proper parts which are disjoint from the lump (the ring, recall, is assumed to be a sum of precisely the same molecules as the lump).

As long as P5 is maintained, therefore, it follows from the denial of P6 that an entity *u* may have an improper part *v* which is not identical to *u*. This, however, contradicts principle P1.

1. Types of Neoclassical Sum

It is an expression of the flexibility and versatility of the notion of a sum, that it is compatible both with the classical theory and with a variety of neoclassical theories. In virtue of this versatility, some of the most important phenomena associated with relations between wholes and their parts can be apparently satisfactorily accounted for in terms of the conception of wholes as sums. Under neoclassical assumptions types of sums can be introduced with features tailored to account for phenomena which were not accounted for satisfactorily in classical mereological terms.[23]

[22] This follows from SCT31 and SCT34 in Simons 1987, 39.

[23] Note that Simons reserves 'sum' for what I have called 'classical sums'. I use the term 'sum' more generally, so that *y* is a sum of the *xs* if and only if it complies with the condition stated in P1 - whether in temporally unmodified or temporally modified terms.

One type is that of sums of which it is an essential characteristic that certain relations hold between their parts, or that the parts have certain properties.[24] I call sums of this type 'conditioned sums'. By contrast, classical mereological sums can be described generally (but not without exceptions) as 'unconditioned sums'.[25] To simplify the discussion, I will consider at present only conditioned sums which involve conditioning *relations* (rather than conditioning properties).

It is convenient to consider such conditioning relations under two headings. One is that of relations such that, if *y* is a sum of the *x*s, the existence of the *x*s does *not* depend on their having those relations to one another. For example, if the *x*s are small pebbles (less than an inch across), the relation of being three inches apart from one another is such a relation. Similarly, the relation of gravitationally attracting one another with a certain force f_0 (in Newtons) is a relation on which the existence of the pebbles does not depend. With certain paradigmatic examples of such relations in mind, examples in which the relations set up the spatial structure according to which the *x*s are arranged, sums which are conditioned by such relations can be called 'structured sums'.

The other type of conditioning relations to consider are relations such that, if *y* is a sum of the *x*s, the existence of at least some of the *x*s *does* depend on the *x*s having those relations to one another. For example, Aristotle famously claims that if a finger is cut of from the hand of a living person, that finger ceases to exist.[26] Therefore the existence of the finger depends, according to Aristotle, on its being appropriately attached to the living person's hand. Sums conditioned by such relations may be called 'internally dependent sums'.

A second type of sums which the neoclassical framework can accommodate is that of sums which are not unique with respect to their parts. That is, under neoclassical assumptions, both *y* and *z* may be sums of the *x*s, and yet *y* is not identical to *z*. Sums with this characteristic may

---

[24] I speak of relations as universals, or as types, and not as universal instances (tropes) or tokens.

[25] Regarding some relations, or some properties, the existence of the *x*s depends on their having those relations to one another, or those properties. Consequently, the existence of a sum of the *x*s depends, *even on classical assumptions*, on the *x*s having those relations to one another, or those properties.

[26] Although a different entity, which looks much the same as the former finger, does now exist, and 'finger' is used homonymously to describe the attached entity and the detached one. See Aristotle *Metaphysics* 1035b23-25.

be called 'non-unique sums', in contrast with the 'unique sums' of classical mereology.

Finally, a third type of sums which can be accommodated within the neoclassical framework is that of sums which have different parts at different times. Working with a temporally modified version of the relation *is a part of*, and treating it within a neoclassical framework (i.e. assuming appropriately adapted versions of P2-P5, but not P1, P6 or P7), sums can be introduced which I describe as 'mereologically varying sums', to be contrasted with the 'mereologically constant sums' of classical mereology.

Summarising the general types of sum which find their place in a neoclassical framework, but not in a classical one, we obtain the following picture:

I Conditioned Sums
Ia Structured Sums
Ib Internally Dependent Sums
II Non-unique Sums
III Mereologically Varying Sums

By distinguishing these as different types of sums, I do not wish to suggest that if a sum is of one of these types it may not also be of one of the other types. On the contrary, it is quite possible that a neoclassical mereologist will combine several of these features in elaborating what he or she takes to be the notion of a sum which offers an adequate theoretical conception of wholes.

For example, a neoclassical theory capturing the views of Chisholm regarding wholes would be one which featured conditioned sums, which are at the same time unique and mereologically constant. The views of van Inwagen, one the other hand, would require sums which are both conditioned and mereologically varying, and yet unique. And finally, Simons' view makes reference to sums which are conditioned, non-unique, and mereologically varying.[27]

[27] See Chisholm 1976, Ch.3 (89ff.); Appendix A (138ff.); van Inwagen 1987; 1991 (first sections to section 9) Simons 1987, section 5.2 (177ff.); Ch. 9 (324ff.).

2. Conditioned Sums and Non-unique Sums

By proposing either conditioned sums or non-unique sums, a neoclassical mereologist offers a way to resolve some of the most pressing problems which face the classical mereologist. The complaint put forward in connection with the Aggregate Argument, that a classical sum survives the dispersal of its parts, while typical wholes do not, does not apply to a conditioned sum - if it is conditioned with respect to relations that are disrupted by the dispersal of the parts. If a tree, for example, is a conditioned sum of its cells, in that its existence depends on the cells being appropriately juxtaposed to one another, then it is no wonder that the tree ceases to exist on the dispersal of its cells, for dispersal of the cells entails that the appropriate relations of juxtaposition cease to hold between the cells, so that the conditioned sum ceases to exist.

Similarly, a neoclassical mereologist may accept the claim that a gold ring and the lump of gold which constitutes it are distinct from one another, while insisting that each of them is a sum of the relevant gold molecules. For by allowing non-unique sums, such a theorist may take both the lump of gold and the ring to be sums of the same molecules, without admitting that these sums are identical to one another.

The means whereby a neoclassical mereologist introduces conditioned sums and non-unique sums are rather straightforward. As regards relationally conditioned sums, one simply rejects the classical principle of Universal Existence of Sums (P7), and introduces in its place qualified principles of the existence of sums. That is, one forgoes the assumption that whenever there exist entities, the *x*s, there exists a sum of those entities. Instead, one assumes that whenever there exist entities, the *x*s, such that the *x*s are conditioned in a particular way (e.g. they bear certain specified relations to one another), there exists a sum of the *x*s.

As regards non-unique sums, one begins by rejecting the classical Principle of Uniqueness of Sums (P6), as well as the principle according to which *is an improper part of* is identified with *is identical to* (P1) (for reasons explained above). Instead, the neoclassical mereologist may rely again on the notion of a conditioned sum, to lay down conditions under which different sums of the same parts exist, where the existence of each of these sums is dependent, respectively, on the preservation of different conditions.

For example, assuming that the *x*s are an appropriately large number of gold molecules, it may be laid down that a sum of these molecules exists, if the molecules are so juxtaposed that they occupy a continuous region of space. This sum may be identified as what we refer to using the term 'lump'. Furthermore, it may be laid down that a sum of the molecules exists, if the molecules are arranged so as to occupy a ring-shaped region of space, and this sum may be identified as what we refer to using the term 'ring'. Clearly, under some conditions, both the former and the latter relations hold between the *x*s. Since, however, under some conditions the former relations hold and the latter do not, the former sum will be claimed to be distinct from the latter sum. Since P6 and P1 are rejected, claiming this is not inconsistent with the neoclassical assumptions.

3. Mereologically Varying Sums

The complaint against classical mereology, that it cannot provide an account of mereologically varying wholes, can be met in neoclassical theory by introducing mereologically varying sums.

Suppose a certain tree has survived a forest fire. However, some of the cells which belonged to it at *t* (before the fire) have been destroyed in the fire. These assumptions are not consistent with taking the tree to be a classical mereological sum. If the sum of the tree's cells at *t* is conceived along classical lines, this sum cannot survive the destruction of any of its parts. The assumptions are consistent, however, with taking the tree to be a mereologically varying sum, one which may have different parts at different times.[28]

The introduction of mereologically varying sums, within a neoclassical framework, is not as straightforward as the introduction of conditioned sums or of non-unique sums. Peter Simons elaborates the notion, or more precisely *several* notions, of mereologically variable sums in Chapter 5 of *Parts*. The first step involves replacing the temporally unmodified mereological predicate 'is a part of' with the temporally modified 'is a part of at *t*', as a primitive of the formal system.[29] Other temporally modified predicates, such as 'is a proper part of at *t*', 'is disjoint from at *t*', and 'overlaps at *t*', are defined in terms of 'is a part of at *t*'.[30] In terms of these predicates, notions of mereologically varying sums can be explicated.

My presentation of the notion of a mereologically varying sum closely follows that of Simons.[31] The main difference is in the emphasis I place

[28] Mereologically varying sums *can*, but need not have different parts at different times.
[29] See Simons 1987, 177-9.
[30] See Simons 1987, 179.
[31] See Simons 1987, 183-6.

on the relation *is a sum-at-*t, which I explain in the next paragraph. Although this relation is only mentioned in passing by Simons, it will be seen that it is central to the various notions of mereologically varying sums which he discusses. Emphasis on the role of this relation allows it to be clearly seen that the complex array of types of mereologically varying sums which Simons presents are closely related to one another, being different versions of one basic type of mereologically varying sums.

I take the notion of a mereologically varying sum to be the notion of an entity which (for some *t*) bears the relation *is a sum-at-*t to some entities.[32] This relation can be defined as follows (note the use of the temporally modified predicate 'overlaps at *t*'):

D3 *y* is a sum-at-*t* of the *x*s $=_{\text{def.}}$
y exists at *t*, and for all *z*, *z* overlaps *y* at *t* if and only if *z* overlaps one of the *x*s at *t*

Let us suppose, for example, that a certain church is built out of bricks, and that bricks are the only entities from which it is built (that is, we are ignoring mortar, beams, etc.). Suppose we treat the church as a mereologically varying sum which exists at *t*, and suppose that bricks a_1, ... , a_n are all the bricks which are parts of the church at *t*. Then the church is a sum-at-*t* of a_1, ... , a_n, for the church exists at *t*, and something overlaps the church at *t* if and only if it overlaps one of a_1, ... , a_n at *t*.

The notion of a mereologically varying sum allows for such a sum to comprise different individuals at different times. That is, if $t \neq t'$, *z* may be a sum-at-*t* of the *x*s, and a sum-at-*t'* of the *y*s, where the *x*s are not (severally) identical to the *y*s. For example, it may be that the church of the above example is a sum-at-*t'* of bricks b_1, ... , b_n, where *t'* is a time later than *t*, and where at least one of these bricks (say, b_i) is not one of a_1, ... , a_n. This would be the case if, for example, fifty years after the church was built repairs were carried out, and several of the original bricks were replaced by new ones.

To specify a mereologically varying sum *y*, one needs to specify, for any time *t*, which *x*s (if any) are such that *y* is a sum-at-*t* of those *x*s. Simons suggests two basic ways in which this might be done:[33]

[32] The notion of a sum-at-*t* presented here corresponds to that of Simons' 'Su_t' sums. See Simons 1987, 184.

[33] See Simons 1987, 183-6.

(1) The first is by means of a list of names. Thus one might speak of a mereologically varying sum of *a* and *b*, where *a* and *b* are individuals which exist over periods of time which may or may not overlap one another ('*a*', '*b*' are the names in the list relevant to this case). Simons considers this way of specifying a sum only with regard to cases involving *two* summands, that is, only with regard to cases where a *binary* sum is involved.[34] He indicates that there are two different types of entity which might be intended by such specification.

He gives one type the name 'SU' sums. An 'SU' sum of *a* and *b* is an entity which exists whenever *either a or b* exists:

D4 *c* is an 'SU' sum of *a* and *b* $=_{\text{def.}}$
for all *t* and for all *x*,
x overlaps *c* at *t* if and only if
x overlaps either *a* or *b* at *t*

He gives the other type the name 'SM' sums. An 'SM' sum of *a* and *b* is an entity which exists whenever *both a and b* exist:

D5 *c* is an 'SM' sum of *a* and *b* $=_{\text{def.}}$
for all *t* and for all *x*,
x overlaps *c* at *t* if and only if
both *a* and *b* exist at *t* and *x* overlaps either *a* or *b* at *t*

If $d_1, \ldots, d_n$ are entities, then an 'SM' sum of those entities is an entity which exists only at such times (if at all) in which every one of the *d*s exists. Clearly, therefore, the notion of an 'SM' sum is not the notion of a mereologically varying sum, for an 'SM' sum may not have different parts at different times (ignoring any mereological variation taking place in any of the *d*s themselves). Thus, although 'SM' sums are explicated in temporally modified terms, they are not mereologically varying sums.

By contrast, an 'SU' sum may well be mereologically varying. In particular, if *c* is an 'SU' sum of $d_1, \ldots, d_n$, such that *c* exists both at *t* and at *t*' (where $t \neq t'$), and if some of the *d*s which exist at *t* do not exist at *t*', then *c* will have parts at *t* which it does not have at *t*'.

---

[34] See Simons 1987, 183-4.

(2) The second way in which a mereologically varying sum might be specified is by means of descriptive phrases, rather than lists of names. Thus one might speak for example of a mereologically varying sum of the schoolteachers, 'schoolteacher' being the relevant descriptive phrase.

Simons shows that the account of sums specified in this way is more complex, because there is a greater scope for ambiguity regarding the type of entity that might plausibly be intended. Simons mentions three such types, although it is clear that more might be distinguished.[35]

One type is that of an entity such that, at any time t at which it exists, it is a sum-at-t of all persons who live at t and who are schoolteachers *at one time or another*. More precisely and generally, taking 'is an F at t' to be a temporally modified descriptive predicate (such as 'is a schoolteacher at t'), this type of sum may be defined as follows:

D6 a is an 'SU' sum of Fs $=_{\text{def.}}$
for all x and for all t
x overlaps a at t if and only if
for some y and for some t', y is an F at t' and x overlaps y at t

A second type is that of an entity such that, at any time t at which it exists, it is a sum-at-t of all persons who live at t and who are schoolteachers *at some specified moment of time t'* (t' might be, for example, the time of an utterance involving the phrase 'sum of the schoolteachers'). Taking '$F_{t'}$s' to designate 'schoolteacher at t'' , this type of sum may be defined as follows:

D7 a is an 'SU' sum of $F\text{s}_{t'}$ $=_{\text{def.}}$
for all x and for all t
x overlaps a at t if and only if
for some y, y is an F at t', and x overlaps y at t

A third type of entity which might be intended by means of 'sum of schoolteachers' is that of an entity such that, at any time t at which it exists, it is a sum-at-t of all persons who live at t and who are schoolteachers *at that time t*. This type may be defined as follows:

---

[35] See Simons 1987, 184-6.

D8 *a* is an 'SUM' sum of *F*s $=_{\text{def.}}$
for all *x* and for all *t*
x overlaps *a* at *t* if and only if
for some *y*, *y* is an *F* at *t*, and *x* overlaps *y* at *t*

Inspection reveals that any one of these three notions of a sum (i.e. any one of the three presented above which involve specification by means of descriptive phrases rather than by means of lists of names) are notions of mereologically varying sums. A sum of schoolteachers, understood as an entity of either of these three types, may clearly be an entity *y* such that, if $t \neq t'$, *y* has parts at *t* which it does not have at *t*'.

This concludes our examination of the various of types of mereologically varying sums, types which are distinguished as being specified either by using lists of names or by using descriptive phrases (interpreted in any one of three ways).

Chapter 6
Traditional Higher-grade Wholes as Sums

A recurrent and striking feature of reflection on wholes and parts, both in ancient and modern times, is the attempt to draw a distinction between what might be described as higher-grade and lower-grade wholes. Perhaps the earliest example of such a distinction is that standing at the source of the present work, namely, Plato's distinction in *Theaetetus* between a complex which is different from all its elements and a complex which is identical to them. A more familiar ancient example is found in Aristotle's distinction between substances and heaps. In early modern times, Leibniz distinguishes similarly between monads and monadic aggregates.[1] In the 19th and early 20th centuries, two notions in particular developed into central representatives of this perennial attempt to explicate a higher-grade type of whole: The notion of an organic whole, and the notion of a Gestalt.[2] The latter two are fairly described as the main traditional modern notions of higher-grade wholes.

Certainly, other terms are often used in modern contexts to designate higher-grade notions. Husserl, for example, uses 'pregnant whole'. Russell uses the term 'unity'. And among thinkers associated with the Berlin Gestalt school we sometimes find 'dynamic whole' and 'functional whole'. More recently, Simons uses 'integral whole', and Lowe uses 'integrate'.[3] It seems to me, however, that in assessing the relations between neoclassical mereological assumptions and the various higher-grade notions, we arrive at representative conclusions by confining our attention to ideas directly associated with the two most familiar terms, that is, 'organic whole', and 'Gestalt'.

The predicate 'is of higher-grade' is used loosely in connection with such notions of wholes, in the following sense: They present the relations between the properties of a whole and the properties of (and relations between) its parts as a more complicated affair than do the contrasting

---

[1] See Plato, *Theaetetus* 201c ff.; Aristotle, *Metaphysics* 1041b11-31; Rescher 1967, Chs. 7 and 9 (regarding monads and monadic aggregates).

[2] For references, see discussion below.

[3] See Husserl 1901, Investigation III, section 22; Russell 1903, 139; Nagel 1952 (regarding dynamic and functional wholes; Simons 1987, Ch.9; Lowe 1989, Ch.6.

lower-grade notions. Motivation for developing higher-grade notions, in contrast with lower-grade ones, can be found in many and varied contexts.

To take one example, minor changes in the relations between the parts of a heap of car-parts are likely to be less consequential for the properties of the heap than comparable changes in the relations between the parts of a car are likely to be for the properties of the car. If one assumes that the difference between the sensitivity of the (properties of the) car and that of the (properties of the) heap to changes in the relations between their respective parts is better viewed not as being a matter of degree, but rather as being a matter of fundamental difference in kind, it would seem plausible to attempt to develop a corresponding theoretical distinction between two notions of a whole, a higher-grade and a lower-grade one.

What I would like to suggest in the current section is that the neoclassical mereological framework is flexible enough to admit sums which have any of the features commonly associated with higher-grade notions of wholes. That is to say, to claim that a whole is a sum is compatible with claiming that it is a higher-grade whole. I wish to discuss this claim in some detail only with respect to two types of higher-grade whole mentioned above, namely, organic wholes and Gestalts.

I am not attempting to establish conclusively that all familiar types higher-grade whole, nor even organic wholes or Gestalts in particular, can be adequately accounted for as types of neoclassical sum. To do so would require a full-length study of its own. My more limited task is merely to indicate that such an account of higher-grade wholes is plausible. This serves two purposes. First, it is meant to show how traditional higher-grade notions fit in from the perspective of the mereologies that we have been considering, and thus to render our theoretical picture less fragmented.

Second, it is meant to persuade us against assuming that familiar higher-grade notions are likely to offer adequate responses to the problems raised (in the next chapter) in criticism of the conception of wholes as sums. If, for example, the principal features by which an organic whole is characterised are fully compatible with treating such a whole as a sum, it should not be expected that the notion of an organic whole could offer an adequate response to criticisms directed at the foundation of the conception of wholes as sums.

I - Organic Wholes and Gestalts

1. Organic Wholes

The notion of an organic whole gained wide currency around the turn of the last century, both in philosophical circles and outside them.[4] Examples of prominent philosophers during this period for whom the notion played a central role include Nietzsche, Bradley, McTaggart and Whitehead. The immediate sources for this notion are often considered to be Goethe, Kant and Hegel, although it seems clear that much older ideas influenced its development, notably ideas due to Aristotle and Leibniz.

Thus James Benziger traces Coleridge's views on organic wholes back to Leibniz, through a complex genealogy which includes Schlegel, Karl-Philip Moritz, Herder and Goethe, and weighs alternative claims which trace Coleridge's views back to Schelling and Kant. And G.E. Moore takes the discussions of organic wholes current in his time, especially in the works of Bradley and McTaggart, to have their principal source in Kant and Hegel. Richard Shusterman claims that Nietzsche's understanding of organic wholes is a radical development of views put forward by Aristotle in the *Poetics*.[5]

Indeed, from the 18th century onward the notion appears to have been an important expression of the reaction against the dominant mechanistic views typical of the Enlightenment. Max Oelschlaeger comments on the role of the cognate notion of an organism as follows:

> Since the publication of Descartes' *Discourse on Method* (1637) and *Meditations* (1641) and Newton's *Principia Mathematica* (1687), the mechanistic model has dominated natural science. The Cartesian-Newtonian paradigm enjoys cognitive hegemony in the modern world, displacing any aesthetic, religious, or philosophical claim to insight or knowledge. ... Modernism draws, perhaps unconsciously but absolutely, a boundary between an objective or scientific and a poetic or aesthetic view of nature.

---

[4] Moore speaks disapprovingly of the widespread, and often very vague, use of the term 'organic whole' in Moore 1903, 27ff. See also Benziger 1951.

[5] See Benziger 1951; Moore 1903, 27ff.; Shusterman 1988. The view that Aristotle's substances should be understood as organic wholes can be found in Bogaard 1979, 12, and similarly in Ross 1924, 219.

However,

Even though Modernism has intellectually ruled 'reason' in the Western world for some four hundred years, challenges to its cognitive adequacy have been numerous. ... Collectively considered, the critics of Modernism engendered an opposition between two rival ideas of nature: *the idea of nature-as-a-machine* as against that of *nature-as-an-organism.*[6]

A central intuition underlying the notion of an organic whole concerns a certain unity* which wholes of some but not all types allegedly possess, a unity* understood somehow as the perceptible presence - equally in each of the many parts of the whole - of one and the same causally responsible item.[7] The type of causation involved here is sometimes assumed to be efficient, sometimes teleological. Living organisms, and particularly plants, are usually taken as paradigmatic wholes of this type - hence the use of the term 'organic'. Benziger, in his discussion of the role of the notion in Coleridge's thought, explains as follows:

The very use of the word *organic* to describe a work of art is a metaphor. It is a comparison of the unity of a work of art to that unity which ordinary men imagine they perceive in a tree, a unity which is the expression of one indwelling force or spirit.[8]

Goethe, for example, says of the Strassburg cathedral that

its great harmonious masses [are] filled with life even to the countless small details, like the works of eternal nature, all form, all contributing to the whole, down even to the most minute fiber.[9]

The unity* of organic wholes is further understood as rendering the parts of such wholes entities which are in some sense not independent, either with respect to one another or with respect to the whole of which they are parts.

---

[6] See Oelschlaeger 1991, 97-98.

[7] I attach an asterix to the term 'unity' here to distinguish it from 'unity' in the sense explained in Chapter 2, sections I and II.

[8] Benziger 1951, 33.

[9] Benziger notes Vol. XXX, 9 of *Goethes Sämtliche Werke*, Jubiläumsausgabe, Stuttgart, 1902-7, as the source for this quotation.

Thus the parts of an organic whole are sometimes taken as having properties which are dependent on the properties of the other parts or of the whole, sometimes as depending for their own existence on the existence of the other parts, or of the whole, and sometimes as incapable of being understood or even conceived independently of one another or of the whole. Furthermore, it is sometimes unclear whether the necessity characterising these dependencies is meant to be causal, ontological or logical. An example in which claims regarding the presence of a single causally responsible item in the many parts are combined with claims regarding the non-independence of those parts can be found in Thomas Wartenberg's account of Hegel's conception of the world as an organism.[10]

As we shall see presently, various features (such as having parts which are functionally or ontologically dependent on one another, or having properties which are emergent with respect to the properties of the parts) have been described in the attempt to explain in more precise terms what the claim that a whole is organic amounts to. It is debatable whether a statement of any or all of those features does justice to the fundamental intuition that gives rise to the notion of an organic whole. My claim that the notion of an organic whole is adequately treated as falling within the framework of the conception of wholes as sums (classical or neoclassical), however, is only made with respect to organic wholes *in as far* as they are taken to be adequately characterised by some or all of those proposed features.

2. Gestalts

The introduction of the notion of a Gestalt is chiefly credited to Christian von Ehrenfels, who discussed it in his article 'Über Gestaltqualitäten'.[11] The notion is particularly familiar in connection with the major influence it has exerted on modern psychology, especially through the work of Max Wertheimer, Wolfgang Köhler and Kurt Koffka. It has also been

[10] See Wartenberg 1993, 102ff., and esp. 108.

[11] Ehrenfels 1890. Ehrenfels takes his views to be a development of some observations made by Mach. Simons, 1987, 356, claims that "the same concept was elaborated and exploited at the time by two other former students of Brentano, Meinong and Husserl, who used the terms 'founded content' and 'figural moment' respectively for what Ehrenfels called a 'Gestalt-quality' ' . For an illuminating account of the development of the notion of a Gestalt, see Smith 1988, 11-81; for a more detailed discussion of Ehrenfels, see Smith 1994, Chs. 8 and 9.

influential, however, in other fields, ranging from the social sciences to artificial intelligence.[12]

In contrast with the case regarding the notion of an organic whole, that of a Gestalt arose not in the context of describing paradigmatic individuals, such as trees or cathedrals, or works of art, such as poems or paintings, but rather in the context of describing types of content of sensation or of perceptual experience. Ehrenfels' discussion begins with a consideration of two examples, that of the shape in which several points are arranged in space, and that of a melody, understood as a musical 'shape' in which several notes are arranged in time.[13]

According to the leading theory of perception at the time (most notably associated with Helmholz), perceptual experience can be adequately analysed into primitive sensations which have qualities such as colour and brightness (in the visual case), or pitch (in the aural case) as their contents.

In the case of visual experience, for example, each point of the visual field is thought of as associated with a primitive sensation - the sensation of a certain colour and a certain brightness at that point. A typical visual experience is viewed as involving many such sensations taking place at the same time (just as the sensation of heat in one finger may be simultaneous with the sensation of coldness in another finger). Furthermore, it is assumed that such a visual experience can simply be *equated* with these many simultaneous primitive visual sensation, which together account for the whole field embraced by the experience.

Suppose, for example, many red dots are arranged in the shape of a circle. According to that theory, experiencing the quality associated with the perception of the circular shape is understood simply as the simultaneous experiencing of the primitive qualities associated with the sensation of each of these dots, respectively.

Similarly, experiencing the quality associated with the perception of a melody is understood simply as the consecutive experiencing of the primitive qualities associated with the sensation of each of the notes, respectively.

---

[12] For a brief survey of ways in which the notion of a Gestalt has featured in a variety of scientific fields, see Palmer and Rock 1990.

[13] A central meaning of the German term 'Gestalt' is 'shape'.

To use Ehrenfels' terms, such an account takes a shape to be a sum* (or a *mere* sum*)[14] of dots, and a melody to be a sum* of notes, and in this respect it seemed to him not true to the familiar experiences of hearing melodies or of seeing shapes. He urged that shapes and melodies, and a great variety of other types of item, be recognised as qualities which, notwithstanding the clear sense in which they depend on the respective primitive qualities, are not to be analysed as sums* of such primitive qualities in the way suggested. He used the term 'Gestalt-quality' to describe the former qualities.

In later usage among Gestalt theorists, the term 'Gestalt' was generally preferred to 'Gestalt-quality'. This is partly to do with the fact that the term became commonly used to designate not qualities of this type, but rather *entities* which possess qualities of this type.[15] In what follows, I shall use 'Gestalt' in the latter sense.

Taking Gestalts and sums* as two types of whole[16], the most noteworthy distinguishing feature of the former in contrast with the latter, in Ehrenfels' view, concerns the way in which the similarity between two Gestalts is not determined, or at least not completely determined, by the similarities between their parts. Let us call this condition, which is met by Gestalts on Ehrenfels' account, 'Ehrenfels' criterion'. This criterion, he claims, is not answered by wholes which are described as sums*. For he takes the similarity between two sums*, by contrast, to be completely determined by the similarities between their parts.[17]

For example, if a sum* of notes has notes as its parts which are respectively perfectly similar to the notes which are the parts of another sum*, then the two sums* are perfectly similar. By contrast, if a Gestalt has notes as its parts (a melody being an example of such a Gestalt) which

---

[14] The term 'sum' ('Summe'), as used by Ehrenfels, is marked with an asterix, to distinguish this use from the one we have adopted, in connection with the treatment of classical and neoclassical mereology.

[15] The ambiguity as between quality and entity possessed of the quality is analogous to the ambiguity in the use of 'structure', either to designate (collectively) the relations between the parts of an entity, or to designate an entity, the parts of which are thus related (in the latter sense, a bridge, for example, can be described as a structure).

[16] For the purpose of discussing the views of Gestalt theorists, we may take 'whole' to have a wider application than we have agreed to adopt otherwise in the present work, so that not only concrete entities, but also shapes and melodies, might be described as wholes.

[17] More precisely, if S is a sum* whose parts are $a_1 \ldots a_n$, and S' is a sum* whose parts are $b_1 \ldots b_n$, then the similarity between S and S' is a function of (and only of) the similarities between a_1 and b_1, ... , and between a_n and b_n.

are respectively perfectly similar to the notes which are parts of another Gestalt, the two Gestalts may still be very dissimilar to one another. The similarity between melodies clearly depends not only on the identity of the notes which are their parts, respectively, but also on the order in which those notes are played. Ehrenfels assumes that the similarity between sums* of notes does not depend on the order of the notes.

We find, then, that there are two aspects to Ehrenfels' insight. One concerns the claim that the similarities between some wholes is not completely determined by the relevant similarities between their parts. Wholes of this type are called Gestalts. The other aspect concerns the claim that wholes of this type can be contrasted with wholes of another type, the similarities between which *are* completely determined by the relevant similarities between their parts. Wholes of the second type are called sums* (or *mere* sums*).

It has been demonstrated by Kurt Grelling and Paul Oppenheim that the first aspect of Ehrenfels' account admits of a satisfactory precise formulation. In their article, 'The Concept of Gestalt in the Light of Modern Logic', they have shown how a feature grounding the similarities between two wholes can be expressed, a feature which is determined not only by the similarities between the relevant parts of the two wholes, but also by the relations between those parts.[18]

It is more difficult to explicate convincingly the second aspect of Ehrenfels' account. The claim that some wholes comply with Ehrenfels' criterion seems to be uncontroversial. What is controversial, however, is the claim that some wholes do *not* comply with that criterion.

For the natural question to ask is, what might count as an example of a mere sum*? Whenever one hears a sequence of notes, these notes are obviously heard in a specific temporal order, and the similarity between this sequence and another will be at least partly determined by the order of the notes in the sequence. And a similar remark would apply to any case in which a whole is seen, the parts of which are dots on a sheet of paper. Turning from the consideration of the contents of sensation and perception to the consideration of concrete entities, again no clear examples of wholes which would be adjudged sums* on Ehrenfels' criterion seem to be forthcoming.

The main response of Gestalt theorists to this difficulty was effectively to abandon Ehrenfels' criterion, while attempting to preserve the intuition

---

[18] See Grelling and Oppenheim 1938. See also Simons 1987, 354-360.

suggested by his seminal essay concerning the contrast between a higher-grade type of wholes and a lower-grade type, which the notion of a Gestalt is meant to explicate.[19] Gestalts are thus taken to be higher-grade wholes, typical examples of which are found in the organic world (e.g. the brain of a human being), but not only in the organic world (e.g. an electrically charged capacitor). An account of the notion of a Gestalt, on this view, would require that the theorist identify particular types of characteristics of a whole, and especially particular types of relations between the parts of a whole, which mark the distinction between wholes that are Gestalts and wholes that are not Gestalts. The characterisations that were actually offered turned out to be, by and large, variations (though often interesting variations) on the familiar characterisations of organic wholes.

Among theorists not belonging to the Gestalt school a different response could be found. They interpreted the notion of a Gestalt not so much as applying to higher-grade wholes, in contrast with lower-grade ones, but rather as applying to any whole whose identity depended on the relations between its parts. On this view, a sand dune, or a heap of old car-parts, are no less Gestalts than is a car, or for that matter an organism. Indeed, Grossmann, for example, claims that any concrete whole is a Gestalt in this sense.[20] He thinks that the only reasonable interpretation for Ehrenfels' sums* is one which takes them to be classes, that is, abstract entities.[21]

Simons also does not consider the notion of a Gestalt to mark a distinction between a whole which would usually be considered to be higher-grade, such as an organism, and a whole which would usually be considered to be lower-grade, such as a sand dune.[22] According to his view, both an organism and a sand dune are Gestalts, although he leaves open the possibility that some wholes exist whose identity does not depend on the relations between their parts, and thus that wholes exist which are not Gestalts. The notion of a classical mereological sum, rather than what I have described as a (neoclassical) conditioned sum, would

---

[19] See, for example, the discussion in Köhler 1929, Ch. 4.

[20] See Grossmann 1983, 242.

[21] See Grossmann 1983, 241. On the plausibility of interpreting Ehrenfels' "sums" as classes, see also Nagel 1952, 143.

[22] Simons does however take it that the notion of a Gestalt, as developed, for example, by Köhler, is valuable in helping to direct us to the important problem, 'what *makes* something a natural (or other) complex, which entails giving a schematic account of the constitutive interrelations of parts characteristic of such complexes.' See Simons 1987, 324-6.

apply to such non-Gestalts, and would provide an adequate characterisation of them.[23]

In what follows, the discussion is confined to the notion of a Gestalt as put forward by the Gestalt theorists. I shall therefore focus on presenting features which have been proposed by Gestalt theorists, features by means of which the notion of a Gestalt is marked out as a notion of a higher-grade whole. Regarding the claim that I argue for below, namely that the notion of a Gestalt is adequately treated as falling within the framework of the conception of wholes and sums, a comment is in place which parallels the one made above with regard to organic wholes: I make this claim only with respect to Gestalts *in so far* as they are taken to be adequately characterised by some or all of those proposed features.

To conclude, it can be seen that notwithstanding their distinct intuitive roots, the notions of an organic whole and of a Gestalt came to be associated interchangeably with one or more features taken from a certain pool of features of higher-grade wholes. I shall now turn to state briefly the main features belonging to this pool, and argue that *sums* may be proposed which have any of these features, thus showing that conceiving of a whole either as an organic whole or as a Gestalt is compatible with conceiving that whole as a sum.

II - Features of Organic Wholes and Gestalts

In *Principia Ethica*, G. E. Moore offers a critique of the notion of an organic whole by gathering what he takes to be the main claims made by the various theorists who utilise the notion, claims that indicate the main features which have been associated with the notion of an organic whole in his view.[24] Thus he says that the description of an entity as an organic whole 'is generally understood to imply' one of the following three claims:

---

[23] Note however that in Simons' usage, by contrast with the one I have adopted (according to which 'sum' applies in general to wholes in classical and neoclassical mereology) 'sum' does not apply to what I have called conditioned sums.

[24] See Moore 1903, 27-36.

(1) The entity has parts which 'are related to one another and to [the whole] itself as means to end'. [25]

(2) The entity is such that its parts 'have a property described in some such phrase as that they have 'no meaning or significance apart from the whole''.[26]

(3) The entity is such that its 'value is different from the sum of the values of [its] parts'. [27]

If we compare Moore's review with those offered by some other commentators on organic wholes and on Gestalts, we find that if these three features are appropriately generalized, they do, by and large, represent the main accounts of the two notions.[28]

1. Causal Dependence between Parts

As regards (1), Moore understands the claim as intended to characterise a feature which living organisms are commonly assumed to possess, whereby the continued existence of some part is a necessary condition for the continued existence of some or all of the other parts.[29] Thus he says that an organic whole in this sense is one in which some of the parts exhibit 'a relation of mutual causal dependence on one another'.[30]

Moore has in mind a relation such that the *existence* of a part causally depends on the existence and proper functioning of another part. More generally, we may speak of relations such that either the existence of a part x, or its having certain properties, or its bearing certain relations (to some of the other parts of the whole), causally depends on the existence of some other part y, or y's having certain properties or certain relations (to some of the other parts of the whole).

---

[25] Moore 1903, 31.
[26] Moore 1903, 31.
[27] Moore 1903, 31; see also 29.
[28] See Nagel 1952; Grossmann 1983, Ch.5; Hartshorn 1942; Wartenberg 1993, Sec.II; Shusterman 1988; Köhler 1929, esp. Ch.4.
[29] In the discussion which follows, 'part' is understood as 'proper part', unless otherwise emphasized.
[30] Moore 1903, 32.

If the properties of, or relations between, the parts of a whole exhibit such causal dependencies, they are said to be *dynamically* determined. Wholes whose parts have properties and stand in relations which are dynamically determined are sometimes described as *functional* wholes.[31]

Moore notes that the relation of means to end might be interpreted otherwise than as such efficient causal dependence. It might, for instance, be interpreted as involving some teleological relation between the behaviour of one part and other parts or the whole.

Indeed, Charles Hartshorn's notion of an organic whole can be stated as precisely such an alternative interpretation of (1). The essential feature of an organic whole according to him is not that the parts of such a whole are causally dependent on one another, but rather that they 'serve as 'organs' or instruments to a purpose or end-value inherent in the whole'.[32] On Hartshorn's account, the criterion for a whole's being organic is that the behaviour of (all) the parts of the whole is *teleologically* related to the existence and behaviour of the whole, and not that it is causally related (in the sense of efficient causation) to the existence and behaviour of the whole or of any of the other parts.

2. Ontological Dependence between Parts

Moore's second sense of 'organic whole', defined by (2), is further elaborated as follows:

It is supposed that just as the whole would not be what it is but for the existence of the parts, so the parts would not be what they are but for the existence of the whole.[33]

He adds that this is distinguished from the feature of causal dependence between parts, expressed in (1) above:

... this is understood to mean not merely that any particular part could not exist unless the others existed too ... , but actually that the part is no

---

[31] See Köhler 1929, Ch. 4, esp. 111-120; See also Nagel 1952, 147, making reference (notes 6 and 7) to Wertheimer, Koffka and Lewin. An interesting formal account of the notion of a functional whole is proposed by Grelling, in Grelling 1987, and discussed in Simons 1987, 342ff.

[32] Hartshorn 1942, 127.

[33] Moore 1903, 33.

distinct object of thought - that the whole, of which it is a part, is in its turn a part of it.[34]

What Moore seems to be saying is that in the case of (2) the existence of the parts is not conceivable except in conjunction with the existence of the other parts (and, presumably, in conjunction with appropriate relations to those other parts). The existence of a part does not merely depend *causally* on the existence of some or all of the other parts, or on their being related in some appropriate way to one another. A stronger kind of necessity links the existence of a part and the existence of the other parts. Shusterman paraphrases Moore's understanding of (2) as follows:

> The reciprocal dependence of the parts is not simply causal but logical. The very essence, features, or character of any part is somehow constituted or modified by its interrelations with other parts in the whole, so that *it would be inconceivable and would make no sense to speak of any part as existing as that same part on its own.* For there is no essence or nature of the part to speak of apart from its interrelations in the whole.[35]

Leaving aside the question of whether Shusterman adequately captures Moore's intentions here, it may be noted that the relations between the parts of an organic whole according to this view belong to types of *ontological dependence.*

In general outline, by saying that entity x is ontologically dependent on entity y we mean that necessarily, if x exists then y exists. According to this definition, relations of causal dependence of the type described by Moore in connection with notion (1) of an organic whole also count as relations of ontological dependence. However, the claim that x is ontologically dependent on y is usually taken to imply that the necessity involved is not merely nomological but rather metaphysical or logical.[36]

In chapter 8 of *Parts*, Peter Simons offers an extensive treatment of ontological dependence, demonstrating that the notion admits of a variety of important distinctions. His treatment is based, as he emphasizes, on

---

[34] Moore 1903, 33.
[35] Shusterman 1988, 382 (italics mine).
[36] See, in this connection, Simons' remarks (in Simons 1987, 295) on the fuzzy boundary between causal and ontological dependence.

Husserl's theory of *foundation* and of *pregnant wholes*.[37] Simons distinguishes between a *weak* notion of ontological dependence, according to which the dependence of an entity on one of its parts counts as ontological dependence, and a *strong* notion, according to which only the dependence on an entity which is not a proper part may count as a relation of ontological dependence.[38] Furthermore, he distinguishes between *rigid* and *generic* dependence. If it is necessary for the existence of an entity *a* that some entity or other of sort *F* exists, then *a* is said to depend *generically* on entities of the sort *F*. If, on the other hand, the existence of *a* necessitates the existence of a *particular* entity *b*, then *a* is said to depend *rigidly* on *b*.

I shall not engage in the present work with the question whether the explication of ontological dependence which Simons offers adequately captures the intentions of Moore, or those of Nietzsche (on Shusterman's account). In Shusterman's case, at least, the view that the parts of a whole are ontologically dependent on one another is associated with a radical reconception of the nature of those parts. A part, it seems to be urged, is to be conceived as *constituted* by its relations with other parts. It is not an entity of the sort I have been talking about, which stands in relations to other entities of this sort; rather, it simply is those relations. And the whole also is not conceived in the familiar way I have been assuming, but rather as a sort of network of relationships.

I am doubtful whether such a metaphysical view can be made sufficiently intelligible, and in any case I will not attempt to elucidate it here. My claims regarding sums whose parts are ontologically dependent on one another are confined to a metaphysical framework which takes the category of concrete entities as basic and not reducible to the category of relations.

---

[37] See Simons 1987, 290-323. Husserl's theory is found chiefly in the *Logical Investigations*, Investigation III. For treatments of Husserl's theory that are alternative to Simons', see Fine 1995.

[38] I am ignoring here a subtlety in Simons' discussion. He defines strong dependence on the basis of a definition of 'absolute independence'. *a* is absolutely independent if it is possible that: *a* exists and that for all *x*, if *x* exists then it is *included* in *x*, where 'included' might be substituted by 'is a part of' (either as governed by the classical assumption P1 or alternatively, in the way Simons suggests on p.112) or by 'is weakly included in' (a notion that arises in Leśniewskian logic, which I do not treat in the present work). Strong dependence can thus be defined somewhat differently, according to the substitution chosen.

3. Emergent Properties of Wholes

Moore's third sense of 'organic whole', associated with claim (3), can be expressed by saying that a whole is organic if its intrinsic value is an emergent property with respect to the properties of its parts, on one or other understanding of the notion of an emergent property. To interpret (3) more precisely, several notions of an emergent property must be distinguished from one another. In characterising these notions, I build on the familiar distinction between determinables and determinates.[39]

Any property which belongs to a particular range of contrary properties is called a 'determinate'. For example, *is red* is a determinate, belonging to the same range of contrary properties as does the determinate *is green*. A range of such contrary properties is often found to be associated with a certain corresponding property which is called a 'determinable', a property possessed by anything which possesses any of the determinates belonging to that range. For example, *is coloured* is a determinable corresponding to the range of determinates which includes *is red* and *is green*. Relations admit of a similar distinction between determinable and determinates. Thus *is 1" away from* and *is 2" away from* are determinates to which the determinable *is at a distance from* corresponds. At present, however, I shall focus on determinable and determinate properties. If F is a determinable, we may speak of the determinates associated with it as 'F-determinates'.

With this explanation of the distinction between determinables and determinates, we may turn to consider different notions of an emergent property:

(a) One way to understand the claim that *has intrinsic value* (as possessed by the whole) is an emergent property is as follows: First of all, one assumes that *has intrinsic value* is a quantitative determinable. A determinable F is *quantitative* if the F-determinates are distinguished from one another as being of a certain degree, a degree which can be fixed numerically, with greater or less precision. Examples of quantitative determinables are: *has weight*, *has volume*, *has electrical charge*. With respect to such determinates a precise notion of the *arithmetical sum* of these properties may be defined: If F_n and F_m are F-determinates, such that F_n is of degree n and F_m is of degree m, then the arithmetical sum of F_n and F_m is F_{n+m}, an F-determinate of degree $n+m$.

---

39 The distinction between determinables and determinates is due to W. E. Johnson (see Johnson 1921, 174). See also Rosenberg 1995; Simons 1987, 343f.

Now, suppose that in a certain case both the whole and each of the parts have a quantitative determinable *F* as one of their properties. Suppose that the whole has *F*-determinate F_n, and that the arithmetical sum of the *F*-determinates of the parts is F_k, where *k* is different from *n*. In such a case we may say that F_n is *quantitatively emergent* with respect to the *F*-determinates of the parts. Let us assume that the same point is expressed by saying that the *F*-ness of the whole is quantitatively emergent with respect to the properties of the parts.

Taken in one way, therefore, Moore's claim (3) can be explicated as the claim that the intrinsic value of the whole is quantitatively emergent with respect to the properties of the parts. This is one sense in which it might be said that *has intrinsic value*, as possessed by the whole, is an emergent property.

(b) A second way of understanding the claim that *has intrinsic value* (as possessed by the whole) is an emergent property is as follows: Suppose a whole has a determinable *F* which *none* of its parts have. In such a case we may say that *F* is *fundamentally emergent* with respect to the properties of the parts. This might be expressed alternatively by saying that the *F*-ness of the whole is fundamentally emergent with respect to the properties of the parts.

Authors who made extensive use of the notion of emergence earlier in this century seem primarily to have had fundamental emergence in mind.[40] Jaegwon Kim summarises the 'doctrine of emergence' associated with those authors thus:

> The doctrine of emergence, in brief, is the claim that when basic physicochemical processes achieve a certain level of complexity of an appropriate kind, genuinely novel characteristics, such as mentality, appear as 'emergent' qualities.[41]

Mentality is not seen as a determinable, determinates of which are possessed both by a complex whole and by each of its parts, but rather as a determinable which is possessed by the whole and yet by none of its parts. This is what Samuel Alexander means when he speaks of mental

[40] The view that wholes may have emergent properties is particularly associated with Samuel Alexander and with C. Lloyd Morgan, in whose views the notion played a central role. Kim mentions in addition G. H. Lewes and C. D. Broad as representatives of emergentist thought. See Kim 1993, 134.

[41] Kim 1993, 134.

processes as having a 'distinctive quality of mind or consciousness', as contrasted with the corresponding neural processes.[42] It is also implicit in C. Lloyd Morgan's claims regarding 'levels' which are discernible in an organism, each level associated with a 'distinguishing quality'. He conceives of an organism as a 'pyramid with ascending levels, and says:

> Near its base is a swarm of atoms with relational structure and the quality we may call atomicity. Above this level, atoms combine to form new units, the distinguishing quality of which is molecularity; higher up, on one line of advance, are, let us say, crystals wherein atoms and molecules are grouped in new relations of which the expression is crystalline form; on another line of advance are organisms with a different kind of natural relations which give the quality of vitality ...[43]

Indeed, the emergentism of Alexander and of Lloyd Morgan involves a stronger claim than that organisms, mineral crystals, and other such entities have properties which are fundamentally emergent with respect to the properties of their parts. Calling the properties with respect to which an emergent property is emergent the '*base* properties', one finds that emergentism is further associated with the following assumptions:[44] (1) that both the entities which have emergent properties and any of their parts are *physical*;[45] (2) that an emergent property is not *reducible* to its base properties; and (3) that an emergent property *supervenes* on its base properties.[46]

The emergentists, therefore, conceived of emergent properties as a particular species of what I am calling fundamentally emergent properties.[47] The claims that I make below regarding the compatibility of

---

[42] See Alexander 1927, Vol. 2, 5-6 (quoted in Kim 1993, 345).

[43] Lloyd Morgan 1923, 35 (quoted in Kim 1993, 345).

[44] See Kim 1993, 344ff.

[45] I do not go into the question what this assumption precisely amounts to. For the emergentists, at any rate, one of its principle expressions was the rejection of Cartesian minds, as well as of Driesch's notion of an entelechy and Bergson's *élan vital*. For a brief discussion of the latter two notions, see Feldman 1995.

[46] Various aspects of reduction and supervenience are discussed in Kim 1993 (see Chs. 4 and 8 in particular).

[47] It seems clear that the claim that a property is fundamentally emergent is consistent with the denial of any, or all, of the assumptions (1), (2) and (3) with respect to it, though I shall leave this point unargued here.

the conception of wholes as sums with the view that a whole has emergent properties is meant to apply to fundamentally emergent properties in general (as well as to emergent properties of the other types that I distinguish), and therefore in particular to emergent properties as conceived by the emergentists.

(c) In addition to the notions of a quantitatively emergent property and a fundamentally emergent property, further notions of an emergent property may be distinguished. Let us say that if a determinable is not quantitative, in the sense explained above, it is qualitative. Examples of qualitative determinables are *has colour*, *has shape*. If F is a qualitative determinable possessed both by the whole and by the parts, then the F-ness of the whole is neither fundamentally emergent, nor quantitatively emergent with respect to the properties of the parts. However, various relations between the F-determinate of the whole and the F-determinates of the parts may be considered as constituting other types of emergence. To take the most obvious possibility, it might be suggested that if the whole has an F-determinate which none of the parts do, then the F-ness of the whole is *qualitatively emergent* with respect to the properties of the parts.

I shall not explore further in this direction, however. For one thing, it seems not to lead to additional plausible interpretations of Moore's claim (3). For another, it does not seem that cases of this type should raise difficulties not already raised by quantitative and fundamental emergence, as regards the question whether wholes which have such emergent properties can be conceived as sums.

In conclusion of this discussion of emergence, I would like to make note of a very weak notion of an emergent property, which I call a 'trivially emergent property': The property of a whole is trivially emergent with respect to the properties of the parts if, simply, it is different from all of the properties of the parts. Clearly, both fundamentally emergent properties and quantitatively emergent properties are in particular trivially emergent. Indeed, sometimes the claim that a property is emergent is understood as the claim that it is trivially emergent in this sense.[48]

[48] Grossmann suggests an understanding of the claim in this way, when he says that 'A structure ... is also more than the class of its parts in that it may have properties which none of its parts has. There are, as it is sometimes put, emergent properties.' (Grossmann 1983, 246), or when he says that a certain 'spatial figure has a certain property, its shape, which

However, it is easy to see that all concrete wholes possess trivially emergent properties (any concrete entity has a weight, for example, that none of its parts do), and so the possession of trivially emergent properties cannot serve as a criterion for distinguishing between higher-grade and lower-grade wholes (i.e. between organic and inorganic wholes, or between Gestalts and sums*, etc.).

To conclude the discussion of the present section to this point - we have noted three general features, by means of which organic wholes and Gestalts have been traditionally characterised:

F1 The whole has parts which are causally dependent on one another (where different types of causal dependence might be intended).

F2 The whole has parts which are ontologically dependent on one another (where this ontological dependence might be characterised in different ways).

F3 The whole has properties which are emergent (either quantitatively or fundamentally) with respect to the properties of the parts.

4. 'The Whole is Greater than the Sum of its Parts'

A feature which has often been proposed as characterising organic wholes and Gestalts is commonly expressed by the slogan 'the whole is greater than the sum of its parts'. Perhaps the most natural interpretation of this slogan is to take it to express a version of the view which has already been noted above, namely that Gestalts, or organic wholes, have properties which are quantitatively emergent with respect to the properties of their parts.[49] In connection with the claim that an *F*-determinate of a whole is quantitatively emergent with respect to the *F*-determinates of the parts, we speak of an arithmetical sum of the *F*-determinates of the parts, and give a clear arithmetical sense to the claim that the *F*-determinate of the whole is different from this arithmetical sum of *F*-determinates of parts. One type of case in which this claim would be true is that in which the *F*-

none of its part has', in virtue of which he takes the shape to be an emergent property (See Grossmann 1983, 249).

[49] Indeed, the slogan interpreted in this sense is suggested by Moore's formulation of claim (3).

determinate of the whole is *greater* than the sum of *F*-determinates of the parts, where again 'greater' has a clear arithmetical sense, namely, 'numerically greater'.

It is perhaps also natural to take the slogan more widely as expressing the view that wholes of the relevant types have properties which are either quantitatively or fundamentally emergent with respect to the properties of their parts. There is a certain analogy between these two notions of emergence, and this analogy may ground a somewhat metaphorical use of the terms 'greater' and 'sum', according to which a whole that has a property which is fundamentally emergent with respect to the properties of the parts can be said, somewhat loosely, to be 'greater than the sum of the parts'.

Grossmann offers an alternative interpretation of the slogan. As we have noted he takes 'sum*' as used by Ehrenfels to designate a class. A concrete whole is clearly different from the class of its parts, and this difference, Grossmann thinks, is what is meant by saying that the whole is *greater* than the sum of its parts.[50] However, the assumption that a whole is greater than the sum of its parts, in this sense, does not offer an additional criterion for counting a whole as being higher-grade. Indeed, this feature follows directly from concreteness, and so is possessed both by lower-grade and by higher-grade wholes.

It is of course hardly resistible in the context of a discussion of mereology to interpret the slogan as expressing a vague dissatisfaction at the suggestion that the nature of wholes is adequately accounted for as classical or even neoclassical mereological sums. For such dissatisfaction to be taken seriously, however, clear reasons must be given for it. The assumption that a whole might have one of the three features, F1, F2 and F3, seems to me to be among the best reasons that can be found in the literature. In the next section, however, I argue that for a whole to possess any or all of these features is fully compatible with conceiving it as a sum (either classical or neoclassical). Thus, there seem to be no good reasons for accepting the claim that a whole is greater than or indeed different from the mereological sum of its parts, as long as this claim is supported only by the assumption that wholes may have one or other of these three features.

---

[50] Grossmann's use of 'sum' seems not to be consistent, though. In the first paragraph of Sec. 101 (Grossmann 1983, 242) he adds that the whole is 'more ... than the sum of its nonrelational and relational parts', and provides an argument for this claim which would be both superfluous and misleading had he used 'sum' here in the sense of 'class'.

III - Organic Wholes and Gestalts as Sums

Having argued that the features by which either organic wholes or Gestalts have been held to be characterised can be summarised as F1, F2 and F3, I now turn to showing that these features are all compatible with the conception of wholes as either classical or neoclassical mereological sums.

I do this by assuming with regard to each of these features that wholes of a certain type *W* (where for '*W*' either 'organic whole' or 'Gestalt' may be substituted) are characterised as having this feature. I then ask whether a whole's being of this type is compatible with its being either a classical sum or a neoclassical sum. Having looked at wholes possessing one of the features in this way, it will be clear how our conclusions can be generalized to positions according to which a whole might have two of the features, or all three of them, and no separate treatment of these cases will be required. For convenience, we may speak of a whole of type *W* simply as a *W*.

1. Sums with Feature F1

Consider first the claim that a *W* is characterised by feature F1. Thus it is claimed that a whole belongs to type *W* if it has parts which are causally dependent on one another. Now this claim is susceptible to two distinct interpretations. It might be interpreted as either

I1 a whole is a *W* if and only if it has this feature at some time *t*

or

I1* a whole is a *W* if and only if it has this feature *essentially*

According to the first interpretation, if a whole has feature F1 at some time *t*, then it is a *W*, even if it does not have the feature at some other time *t*' (and even if it *might* not have the feature at *t* or *t*'). According to the second interpretation, a whole is a *W* only if it has, and moreover must have, this feature at all times at which it exists. Clearly, if a whole is a *W* on interpretation I1*, then it is also a *W* on interpretation I1.

I am not aware of any explicit statement, either on the part of authors expounding the notion of an organic whole or on the part of those expounding the notion of a Gestalt, which would indicate whether they have I1 or I1* in mind. I1* would seem to be the more natural one. On either interpretation, however, the possession of this feature by a whole is compatible with its being a sum. For on interpretation I1, both a classical sum and a neoclassical sum might be *W*s. While on interpretation I1*, although a classical sum cannot be a *W*, a neoclassical sum can. We can see this as follows.

According to interpretation I1, a classical mereological sum might be a *W*. For suppose many entities exist which are causally related to each other at time *t*. Classical mereological assumptions guarantee that a sum of those entities exists. This sum is an entity which has feature F1 at *t*, and therefore according to I1 it is a *W*. If at some other time *t*' the entities disperse and cease to be causally related to one another then their sum still exists. At *t*' therefore the sum will not have feature F1. This however does not conflict with the sum's being a *W* according to I1. According to I1, then, the possession of feature F1 by a whole is compatible with that whole's being a classical mereological sum. And since the neoclassical assumptions are weaker than the classical ones, the possession of F1 by a whole according to I1 is *a fortiori* compatible with its being a neoclassical sum.

According to interpretation I1*, a classical sum cannot be a *W*, if we assume that the parts of this sum might disperse and cease to be causally related to one another, and yet continue to exist. However, a *neoclassical* sum may be a *W* according to this interpretation. For suppose again that many entities exist which are causally related to one another at time *t*. It is consistent with neoclassical assumptions (though not with classical assumptions) that these entities have a sum only if they are thus related to each other. Such a sum is described as a conditioned sum. It seems correct to say that the conditioned sum is essentially conditioned as it is. Therefore, the sum is a *W* according to I1*. The possession of feature F1 by a whole, according to I1*, is thus compatible with the whole's being a neoclassical mereological sum.

2. Sums with Feature F2

Let us now consider the claim that a *W* is characterised by feature F2. In this case it is claimed that a whole belongs to type *W* if it has parts which are ontologically dependent on one another. Again the claim admits of two interpretations: Either

I2 a whole is a *W* if and only if it has this feature at some time *t*

or

I2* a whole is a *W* if and only if it has this feature essentially

As before, I will not attempt to determine which of these interpretations better represents the views of the relevant organic whole theorists or Gestalt theorists. It is sufficient for us to see that on either interpretation the possession of F2 by a whole is compatible with its being a sum. Indeed, on either interpretation the possession of F2 by a whole is compatible both with its being a classical sum and with its being a neoclassical sum. We can see this as follows.

Suppose the *x*s are entities which are all ontologically dependent on one another. On classical assumptions, there exists *y* which is a sum of the *x*s. Moreover, on classical assumptions, any sum has the parts it has essentially. Thus *y* is such as to have, essentially, parts which are ontologically dependent on one another. Therefore, on interpretation I2*, *y* possesses feature F2, and yet *y* is a classical sum. On interpretation I2*, then, the possession of F2 by a whole is compatible with its being a classical sum. And this implies compatibility (regarding classical sums) on interpretation I2 as well.

The conclusion of the last paragraph implies also that the possession of F2 by a whole is compatible with its being a neoclassical sum, on either interpretation, because of the weakness of the neoclassical assumptions relative to the classical ones.

3. Sums with Feature F3

Finally, let us turn to the claim that a *W* is characterised by feature F3. According to this claim, a whole belongs to type *W* if it has properties which are emergent (either quantitatively or fundamentally) with respect to the properties of the parts. The claim might be interpreted as being either

I3 a whole is a *W* if and only if it has this feature at some time *t*

or

I3* a whole is a *W* if and only if it has this feature essentially

As before, leaving aside the question as to which of the interpretations is historically more faithful to the views of theorists who propounded the notions of an organic whole or of a Gestalt, we find that on either interpretation the possession of F3 by a whole is compatible its being a sum. In particular, we find that on interpretation I3 both classical and neoclassical sums may be *W*s, and that on interpretation I3* a neoclassical sum may be a *W*, although a classical sum cannot be one.

Classical mereology, in assuming that for any *x*s there exists a *y* which is the sum of those *x*s, places no explicit constraints on the properties which *y* may have, given the properties of the *x*s. Implicitly, of course, there are clear constraints with regard to one type of properties which *y* has, properties which may designated as 'mereological'. For example, if one of the *x*s is green, then the mereological assumptions imply that *y* has the property *has a green part*. It seems that apart from such mereological properties, however, no constraints applying to the properties of a sum vis-à-vis the properties of its parts can be derived from the mereological assumptions.

For example, if *y* is a sum of three parts, each of which weighs one kilogram, there is nothing in the classical assumptions to rule out the possibility of *y*'s weighing fifty tons. And similarly, if *y* is a sum of so and so many trillions of molecules, none of which possess the property of consciousness, the classical assumptions are silent as to whether *y* possesses this property or not. If the sum does possess the property of consciousness at time *t*, then it possesses feature F3, and therefore is a *W*, according to interpretation I3. And since being a *W*, on I3, is compatible with being a classical sum, it is also compatible with being a neoclassical sum.

However, it is highly plausible to assume that if the sum of those molecules has the property of consciousness at time *t*, then its having this property depends (for example) on the molecules being spatially related to one another in some particular way.[51] If at some other time *t*' each of the molecules is thousands of kilometres away from any of the other molecules, then it would seem very unlikely (to say the least) that *y* could have the property of consciousness at *t*'. And yet according to classical assumptions, as long as the molecules exist, whether arranged as the parts of a living organism or not, their sum *y* exists. On plausible (non-

[51] This assumption is weaker than the assumption that the emergent property supervenes on the properties of and relations between the parts. To say that it supervenes on those properties and relations is to say that the sum has this property *if and only if* the parts have those properties and are thus related to one another. While I am merely assuming on that the sum has this property *only if* the parts have those properties and are thus related to one another.

mereological) assumptions, therefore, it seems that although a classical sum may, on interpretation I3, have feature F3, it cannot on interpretation I3*.

Neoclassical assumptions however allow us to propose, for example, that a sum of the molecules exists only if those molecules are arranged in a particular way (this would be a conditioned sum). It would be an essential feature of such a sum that its parts are molecules which are arranged in that particular way. If this condition regarding the molecules is sufficient for the sum to have the property of consciousness, then such a sum would have the property of consciousness essentially, and thus would have feature F3 and would be a *W*, according to interpretation I3*.

That a neoclassical sum might be a *W* according to interpretation I3* may be demonstrated more directly however. The condition associated with a conditioned sum may pertain not only to relations between the parts, but also to properties of the parts. In particular, since according to neoclassical assumptions (as well as classical assumptions) a whole is a part of itself, the notion of a conditioned sum includes that of a sum which possesses a certain property essentially. In particular, it may be assumed that a sum possesses an emergent property essentially. For example, it is consistent with neoclassical assumptions to assume that a sum exists which essentially possesses the property *is conscious*. This shows that a whole's being a *W*, on interpretation I3*, is compatible with its being a neoclassical sum.

We may conclude that according to a variety of familiar accounts of the notions of an organic whole and a Gestalt, the claims that the whole is an organic whole or that the whole is a Gestalt are compatible with the claim that the whole is a neoclassical sum, and sometimes even with the claim that it is a classical sum. Our discussion suggests, therefore, that conception of wholes as sum is flexible enough to accommodate traditional modern conceptions of higher-grade wholes, and is fully compatible with them.

Chapter 7
Criticism of the Notion of a Neoclassical Sum

Having looked at the advantages which neoclassical mereology offers as contrasted with its classical ancestor, in allowing for conditioned sums, non-unique sums, and mereologically varying sums, and in providing what might seem to be an adequate framework for accounting for traditional higher-grade notions of wholes, I now turn to consider some weaknesses of this flexible theory.

It may be noted, in the first place, that a classical theorist will be unhappy with neoclassical mereology, if she takes it to be one of the main objectives of mereology to provide an alternative to the theory of classes (or set-theory) for the purpose of analysing statements of common discourse. For example, Goodman proposes to analyse

Every species of dog is exhibited
as
For every *x*, if *x* is a species of dog then some *y* is a dog and is part of *x* and is exhibited

where he says that a species of dog 'may be regarded as a certain discontinuous whole composed of dogs',[1] instead of as a class whose members are dogs of that species. Without the principle of Universal Existence of Sums (P7), Goodman would not be justified in assuming that the species, understood in this sense, exists.[2] With Leśniewski, as with Goodman, mereology originated as an attempt to do away with ontological commitment to classes, so that adherence to the classical principles is essential to their project. It is not, however, essential to the different project of accounting for the pre-theoretical notions of whole and part, and someone who acknowledged the existence of classes would see little point in constraining mereology so as to serve the logical roles that Leśniewski and Goodman require.[3]

---

[1] See Goodman 1951,29.
[2] For further discussion of the project of analyzing common discourse by mereological means, see Goodman 1951, 26-33.
[3] See on this point Simons 1987, 101-105.

Granted that ours is the latter project, reflection on the conception of wholes as sums, which neoclassical mereology proposes, leads us nevertheless to discover a multitude of inadequacies in this conception, and suggests that the elaboration of an alternative conception of wholes is desirable, at least in connection with some of the important contexts in which the notion of a whole features.

In what follows I discuss some of the problematic aspects of the three main types of neoclassical sum. Although the problems I point to do seem to me to be serious, and although I doubt whether they could be adequately resolved within the framework of the conception of wholes as sums, I do not take the discussion to constitute a *refutation* of the neoclassical position. To attempt to offer such a refutation would require a much lengthier discussion, and would in any case not serve my purpose.

The main interest in discussing these problems, from my point of view, is to provide an instructive backdrop against which to evaluate the alternative conception of wholes as Unities which I present in the next part of this book. Thus I have attempted to focus on particular aspects of the conception of wholes as sums which give rise to (what I take to be) substantial difficulties, aspects with respect to which (as we shall see) the conception of wholes as Unities differs. Because it differs in this way from the conception of wholes as sums, the difficulties to which I point here either do not arise at all in the context of a theory of Unities, or if they do arise, they have straightforward solutions.

I - Conditioned Sums

In classical mereology it is assumed that for any *x*s there is a *y* such that *y* is a sum of the *x*s (P7). Denying this assumption in neoclassical mereology allows one to introduce the notion of a conditioned sum. This is done by assuming that for any *x*s, there is a *y* such that *y* is a sum of the *x*s if and only if the *x*s fulfil a certain condition (e.g. a certain relation obtains between the *x*s).

At first sight, this neoclassical proposal is very attractive. It seems that wholes are often structured entities. At least in the case of what I have called *prima facie* individuals, the parts of the whole are related to one another in such a way that they exhibit collectively a certain structure. The existence of the whole in such cases clearly depends on the preservation of that structure, in the following sense: The whole cannot be made up of those parts unless the relations between them comply with certain constraints. The introduction of conditioned sums seems to be a promising way in which to express this observation about wholes, for the existence of such sums depends precisely on the preservation of some structure which is exhibited by their parts.

A closer look at the way structured wholes are accounted for as sums, however, reveals a peculiar problem. One seems bound to render the conditions under which parts have a sum as strikingly, and implausibly, *sensitive*. The neoclassical theorist is oddly without freedom to impose conditions as she deems appropriate in this respect.

Let us imagine a human organism h made up at t of a great many molecules $m_1 \dots m_n$ (I shall take 'the *m*s' to abbreviate '$m_1 \dots m_n$'). It would seem that these molecules, considered collectively, are capable of making up a whole of a certain type, when they are appropriately related to one another. If the relation *make up* is interpreted as the relation *compose*, as it is both in classical and neoclassical mereology, then we should say that the *m*s are capable of composing h when they are appropriately related to one another. And to say that they compose h is equivalent to saying that h is their sum.

Now suppose an additional molecule m^+ (not identical to m_1 or ... or m_n) is juxtaposed to the *m*s without otherwise changing the conditions fulfilled by the latter molecules, and suppose this juxtaposition results in m^+ being an additional part of h. Noting that the *x*s compose u if and only if u is a sum of the *x*s, and recalling the definition of a sum, whereby the claim that u is the sum of the *x*s is equivalent to the claim that for all z, z overlaps u if and only if z overlaps one of the *x*s, we directly conclude that after the addition of the extra molecule the original molecules can no longer compose the organism. For if we assume that the *m*s and m^+ (together) compose h, then it follows that the *m*s do *not* compose h. Again, if we assume that to make up is the same as to compose, this means that the original molecules, $m_1 \dots m_n$, no longer make up h.

This seems to be quite a remarkable predicament. Countless trillions of molecules, taken collectively, possess a noteworthy capacity to make up a human organism, if appropriately conditioned, and yet this capacity can be disrupted by the juxtaposition of the most insignificant, virtually undetectable particle. The many molecules, it turns out, instantaneously cease to make the organism up the moment the additional molecule has become a part of the organism.

When m^+ becomes a part of h, the multitude of molecules *ms* would seem to undergo one of the following two dramatic changes, even though we assume that neither their properties (e.g. the mass, charge, and kinetic energy of each of the molecules), nor the relations between them (e.g. their relative positions or motions, the forces by which each attracts or repels the other, etc.), change in any way. Possibly, (1) those molecules cease to make up an entity altogether. This would be consistent with neoclassical assumptions, since they do not include the principle of Universal Existence of Sums (P7).[4] Alternatively, (2) those molecules change from making up a human organism to making up an entity of a radically different kind. This entity, which would clearly be a proper part of *h*, could hardly be an organism of any kind, let alone a *human* organism (interesting exceptions aside, proper parts of organisms are generally not themselves organisms).[5]

Note also that although I assumed that the radical change in the collective ability of the original molecules was engendered by a newly added molecule, it might just as well have been an electron, or a quark. And similarly note that an analogous argument could be formulated taking a far larger entity than a human organism - say, a whale, or perhaps even a planet or a galaxy (if these are accepted as *bona fide* entities).

The assumption that an entity is a conditioned sum seems to imply, therefore, that the collective capacity of a great number of particles to make up an entity would be radically disrupted in the presence of an additional particle. The existence of a sum of the original particles depends not only (plausibly) on relations between those particles, but also (implausibly) on whether or not an additional single particle is present in the vicinity. To put it otherwise - before m^+ was added, the *ms* possessed the following collective attribute: *are such as to make up a human organism.*[6] The approach of m^+ caused the other molecules to cease to have this attribute. On the face of it, it seems surprising that one molecule should have such an effect on so many molecules.

---

[4] That this is what happens in such a case is van Inwagen's view (1981).

[5] See in this connection Simons' criticism of Chisholm's assumption that a proper part of a table may be a table (Simons 1987, 192-3). Simons notes that his point had already been made by Wiggins and by Quine.

[6] Such attributes are described as mutigrade relations. See appendix to Chapter 2 above.

It might be responded that the impression that a single molecule exerts a disproportionate causal influence, assuming that the organism is a conditioned sum, is the result of a misleading description of the circumstances we are considering. Granted, the approach of the new molecule results in the original molecules' ceasing to make up the organism. However, they still have an attribute after the addition of the new molecule which is very much like the attribute they had before. Before they had the attribute *are such as to make up a human organism.* Afterwards they had the attribute *are such as to contribute overwhelmingly to making up a human organism.* These attributes, it is suggested according to this response, are very similar - indeed, virtually indiscernible, precisely in accordance with the limited influence of a single new molecule.

I don't think that this response is satisfactory, however. If the *m*s (conditioned appropriately) were such as *not* to make up a human organism, but the *m*s (identically conditioned) together with m^+ *would* make up a human organism, then it would seem appropriate to say that the *m*s 'contribute' to making up a human organism. Indeed, in such a case the addition of m^+ does not change the attributes of the *m*s in any radical way. For the *m*s were neither making up *h* before the addition of m^+, nor were they doing so afterwards. Both before and after the addition they could be described as contributing towards making up *h*. Before, this was without result, because an additional required contribution, however small, was lacking. Afterwards, the contribution was effective in co-operation with the contribution from molecule m^+.

The case which we have been examining, however, is different. In our case, the *m*s (conditioned appropriately) alone *do* make up a human organism, just as much as the *m*s (identically conditioned) together with m^+ would do. The following analogy may illuminate the relationship between these two types of cases.

Let us assume that if a single bullet pierces the heart of a person - no matter where in the heart - that person dies. Let us also assume that anyone who shoots a person with the result that a bullet pierces the heart of that person, can be said to have caused the person's death. Now, if two gunmen shoot the person so that two bullets pierce the person's heart simultaneously, it would not be appropriate to say of one of the gunmen that he merely *contributed* to causing the person's death. Both, it would seem, are fully responsible causally (we might perhaps say 'overlappingly' responsible).

Developing the example, let us assume that if a hundred bullets or more simultaneously pierce the heart of a giant, the giant dies, and those who shot the hundred bullets (where each gunman is assumed to fire no more than one bullet) have collectively caused the giant's death. If less than a hundred bullets do so, the giant does not die. If one hundred and one bullets are shot, and simultaneously pierce the heart of the giant, then it would be (in parallel to the case of the last paragraph) inappropriate to say of a hundred gunmen that they (collectively) *contributed* to the giant's death. Any hundred gunmen (among the hundred and one) can be said to be (collectively) fully responsible for the giant's death. Any ninety nine or less (among the hundred and one), by contrast, *can* be said to have merely contributed to the giant's death.

Our case of the molecules and the human organism, after the addition of a molecule, is somewhat similar to the case of the giant and the hundred and one bullets. Just as any hundred gunmen - taken collectively - are *fully* responsible causally for, and do not merely contribute to, the giant's death, so the molecules, the *m*s, seem to be fully responsible in respect of making up a human organism, and do not merely contribute to making up such an organism.

I conclude that the response to the problem I have raised is not satisfactory. It fails to pay attention to the distinction between a case in which many molecules (taken collectively) merely contribute to making up an organism, and a case in which many molecules (taken collectively) are fully responsible, in respect of making up the organism. The neoclassical conception of a conditioned sum allows that the original molecules have such full responsibility in respect of making up the organism only before the new molecule was added, and not after. In this way it compels us - if we acknowledge the difference between merely contributing-to-making-up, on the one hand, and fully making up, on the other hand - to accept that the addition of a molecule has a totally disproportionate influence on the attributes of the original molecules taken collectively. I take this to be an indicative weakness of the proposal to conceive of organisms (for example) as conditioned sums.

II - Non-unique Sums

It is among the advantages of neoclassical mereology that it allows for the same entities to have more than one sum. As I have noted in section III of Chapter 5, the rejection of principles P6 and P1 has the result that it is consistent with the remaining principles that for some *y*, *z*, and *x*s, *y* is a sum of the *x*s and *z* is a sum of the *x*s and yet *y* is not identical to *z*. That is to say, it is neoclassically consistent to assume that two distinct entities are superposed with one another, and in particular that two distinct entities are at the same place at the same time.[7]

The flexibility of the neoclassical framework in this respect is an advantage particularly because there are compelling reasons to think that distinct wholes may be superposed with one another at various stages of their career. Recognition of these reasons is largely due to David Wiggins, and the wide acceptance of the possibility (indeed the prevalence) of distinct superposed wholes, following Wiggins' work, is indicated by Burke's description of the view as 'the standard account'.[8] As Wiggins points out, however, the view (or a version of it) is not new. John Locke, for example, accepted that wholes which are not of the same kind (e.g. a 'mass of matter' and a horse which it constitutes at some time *t*) may be superposed with one another (he speaks of being in one and the same place and time, rather than of the sharing of parts; though the latter seems to have been implied).[9]

If (as it would seem difficult to deny) a bronze statue does not exist after the lump of bronze it is made of has been reshaped into a bronze ball, and if (again, as it would be difficult to deny) the lump of bronze exists both before and after this reshaping, then on Leibniz's principle of

---

[7] I have noted above that 'superposition' is defined so that the claim that *u* and *v* are superposed with one another is equivalent to the claim that for some *x*s, *u* is a sum of the *x*s and *v* is a sum of the *x*s. By saying that *u* and *v* are at the same place at the same time, I mean that they are contained in precisely the same region of space at the same time, without implying that they share parts at all (and thus without implying that they are superposed with one another). Arguably, a mind might be in the same place at the same time as a body, without their sharing any parts, and in particular without their being superposed with one another. See Meirav 2000b.

[8] See Wiggins 1968; 1980; Burke 1992, 12 (n.1).

[9] See Locke 1689, Ch. XXVII, esp. 330-331.

indiscernibility of identicals it would follow that the statue and the lump are not identical to one another. For after the reshaping it is the case that the lump is spherical, but it is not the case that the statue is spherical.[10]

In order to resist this inference, one would need to resort to one of several possible extreme measures, such as that of denying that statues or lumps (or both) exist,[11] or of denying that entities might be identical across time,[12] or of restricting in one way or another the scope of Leibniz's principle.[13] It must be admitted however that there is (at least initially) a strong intuitive inclination to resist this inference.

I will argue below that while the assumption that distinct entities may be superposed is not in itself objectionable, the assumption that distinct *sums* may be superposed is objectionable, for if two sums are superposed, they are indiscernible in a certain respect which renders the claim that they are distinct incoherent. Thus while there are good reasons to think that there are non-unique wholes (i.e. distinct wholes which are made up, at one and the same time, of exactly the same parts), there are also good reasons to think that there cannot be non-unique sums. My conclusion from this is that if the arguments of Wiggins and others concerning the possibility of superposition are accepted, an alternative must be found to the conception of wholes as sums.

Before presenting my objections to non-unique sums, however, I wish to consider an argument of David Lewis's which might be taken to constitute an objection to the claim there can be non-unique *wholes*, conceived as sums or otherwise. It will be useful to see why his stronger

---

[10] An elegant version (illustrated by different examples) of the argument briefly presented in this paragraph can be found in Wiggins 1968.

[11] On the view that Doepke calls 'The One-Many View' the lump (in this example) does not exist, strictly speaking. 'Lump' would be interpreted as designating the many bronze particles. On the view he calls 'The Reductivist View' neither the lump nor the statue exist, strictly speaking. See Doepke 1982.

[12] This would still leave the objector to superposition, such as Lewis or Noonan, with an unresolved parallel problem that can be posed with regard to modal properties, involving properties that the statue and the lump *might* have had at the time in question (rather than properties that they actually do have at times different from the time in question). For answers the objectors might offer to this problem, as well as criticisms of these answers, see e.g. Johnston 1992, Noonan 1993 (and references therein).

[13] Doepke 1982, 47-50, argues that avoiding the consequence of superposition by asserting that identity is relative to sort or to time amounts to an arbitrary restriction of Leibniz's principle. The view that identity is relative to sort is developed by Peter Geach (1980), and by Nicholas Griffin(1977). For criticisms of this view, see especially Wiggins 1980, Ch.1, and Lowe 1989, Ch.4., as well as Doepke 1982.

conclusion is not convincing, before arguing for the weaker conclusion that I propose.

1. Lewis's Objection to 'Unmereological Composition'

Lewis, in his article 'Against Structural Universals',[14] makes a point which if acceptable would constitute an objection to the suggestion that there might be non-unique wholes. He comments as follows:

> ... how can two different things be composed of exactly the same parts? I know how two things can be made of parts that are qualitatively just the same - that is no problem - but this time, the two things are supposed to be made not of duplicate parts, but of numerically identical parts. That, I submit, is unintelligible.[15]

Lewis's discussion suggests that the claim

(1) y and z (where $y \neq z$) can both be composed of exactly the same *x*s

can be understood in two ways, but that on either reading it is unacceptable. On one reading the term 'compose' (and cognates) is understood literally; on the other, it is understood metaphorically. Now, it seems to me that there is nothing in what Lewis says in criticism of the literal reading of (1) which indicates that this claim is *unintelligible*. Of the points he makes, the only ones that could be taken as an explanation of why he takes (1) to be unintelligible are points that he makes in connection with the metaphorical reading of (1). However that may be, let us see how Lewis argues for the unacceptability of (1) on either reading, apart from the bald statement quoted above regarding its unintelligibility.

When he speaks of a *metaphorical* understanding of 'the *x*s compose *y*', Lewis has in mind a possible extension of the application of such statements to cases in which the relation between any of the *x*s and *y* is *similar* to the relation *is a part of*. If I were to describe the relation between my parents and me, for example, as a kind of composition, I would be using 'compose' in such a metaphorically extended way. The relation *is a parent of* is similar in some formal respects, but not in others,

[14] Lewis 1986.
[15] Lewis 1986, 36.

to the relation *is a (proper) part of* (for example, both relations are transitive).[16]

Lewis's basic argument is that if one asserts (1) using the metaphorical sense of 'compose' (denying (1) if the literal sense is used), one is asserting something that may well be true, but something that is in conflict with the further claim that the *xs* are *parts* of either *y* or *z*. For unless the *xs* compose *y* in the literal sense, there is no justification for claiming that the existence of *y necessarily entails* the existence of the *xs*. However, according to Lewis, the claim that the *xs* are *parts* of *y* implies that there *is* such a necessary connection between the *xs* and *y*. It follows that to use 'compose' metaphorically and yet to insist on a necessary connection between the *xs* and *y* is to admit a profound modal mystery into one's view.[17]

If on the other hand one asserts (1) using the literal sense of 'compose', then one is asserting something that is simply false, because it is inconsistent with what Lewis takes to be a *the* uncontroversial account of the (literal) notion of composition. This account is of course the one offered by classical mereology, the only theory to which Lewis seems to think the term 'mereology' appropriately applies.[18] Indeed he assumes that if composition in the literal sense were to be considered as consistent with (1), then it should be described as 'unmereological' composition.[19] He denies, however, that a coherent notion of such 'unmereological' composition is available. If it were available, then (classical) mereological composition would be a special case of some more general notion. But he denies that there is a more general notion than the classical mereological one:

---

[16] See Lewis 1986, 38.

[17] Lewis describes the conception according to which a structural universal is composed of other universals (the instantiation of which is necessarily entailed by the instantiation of the former) - in the case that 'compose' is used in a metaphorical sense - as 'the magical conception', referring to the inexplicability of the asserted necessary connection. See Lewis 1986, 41f.

[18] Indeed, in *Parts of Classes* he offers a detailed defence of the principles of classical mereology. See Lewis 1991, 75-81. Comments on Lewis's attitude to mereology can be found in Simons 1991. For criticism of the supposed privileged status of classical mereology, see Doepke's comments in Doepke 1991, 393, as well as Simons 1987 (in general).

[19] David Armstrong, whose notion of a structural universal is criticised by Lewis in the paper we are discussing, also uses of the term 'mereological' in the sense of 'classical mereological'. See Armstrong 1991, 190.

What is the *general* notion of composition, of which the [classical] mereological form is supposed to be only a special case? I would have thought that [classical] mereology already describes composition in full generality.[20]

There is however no reason for neoclassical mereologists to concede that they are using 'compose' metaphorically, rather than literally. Lewis seems to assume that the account of composition offered by classical mereology is the only coherent account available. This, however, is precisely what the neoclassical theorist denies. The neoclassical theorist not only claims that there can be an alternative account, but goes ahead to offer one, in the shape of the neoclassical principles and various assumptions which are formally consistent with them regarding the existence of conditioned sums, non-unique sums and mereologically varying sums.

To the claim that they are using 'compose' metaphorically, neoclassical theorists might respond by pointing out that their account excludes such relations as *is a parent of* from counting as species of the relation *is a part of*, no less than the classical account does.[21] It seems fair to say, therefore, that the contrast between classical and neoclassical accounts of composition is not a contrast between accounts of literal and metaphorical senses of 'compose', but rather a contrast between different accounts of the literal sense of the term.

Lewis might object that sums conceived in accordance with neoclassical principles make modal connections between wholes and parts mysterious. Indeed, in as much as neoclassical principles allow a whole to lose or gain parts, they imply that the existence of a whole y at t does *not* necessarily entail the existence of all its parts at t. This however is not to make admitted modal connections mysterious, but rather to refuse to admit modal connections which would rule out mereological variation. Whichever modal connections the neoclassical theorist *does* acknowledge will be reflected in her version of neoclassical mereology, it seems, no less than the classical theorist's will be reflected in his version of classical mereology.

---

[20] Lewis 196, 39. Additions in square brackets mine.

[21] Note, for example, that Principle P5, of those that I listed, does not apply to the relation *is a parent of*.

Lewis gives us no reason for investing classical mereology with the status of an unquestionable authority on the literal notion of composition. As regards internal consistency, the principles of neoclassical mereology fare as well as those of classical mereology (in fact, since the former is the weaker system, its inconsistency would imply the inconsistency of the latter). And Lewis says nothing which would lead us to think that the assumption that non-unique sums exist is in any way in conflict with those principles. In particular, he does not explain what it is about the assumption that the same *x*s can simultaneously compose (in the literal sense) different entities which supposedly renders it unintelligible.

We conclude that Lewis's discussion does not cast a compelling doubt on the intelligibility of the notion of a non-unique sum, and thus does not provide good reasons for rejecting it.

2. Conflict between Non-identity and Present Indiscernibility

There are three aspects to the state of the allegedly superposed statue and lump that one might find objectionable. One is that the two entities are supposed to occupy precisely the same region of space at the same time. A second aspect is that there are some entities, the *x*s, of which both the statue and the lump are supposed to be sums. And a third is that the two entities are supposed to be *presently indiscernible* at the time in question.

Two entities *x* and *y* are said to be presently indiscernible at some time *t* if they share all their *present* properties at *t* (if either *x* or *y* has a property at *t* which the other does not have, then they are said to be presently discernible at *t*). By describing a property of *x* or of *y* as present at *t* I mean that it is both *modally blinkered* and *temporally blinkered*. A property *F* is temporally blinkered if its instantiation at *t* does not entail the instantiation of any property at any time other than *t*. A property is modally blinkered if its instantiation in the actual world does not entail the instantiation of any property in any world other than the actual one. Properties that are either not modally blinkered or not temporally blinkered are *absent* properties.

Simons takes it for granted that superposed pairs such as the statue and the lump are presently indiscernible.[22] It seems to me, however, that the conclusion that they are presently indiscernible - assuming that they are made up of precisely the same entities - is inescapable only if the statue and the lump are conceived (as by Simons) as sums. To see this, let us

[22] See Simons 1987, 211; 213.

consider on what plausible grounds it might be argued that two superposed wholes are *not* presently indiscernible, suspending briefly the assumption that those wholes are to be conceived according to neoclassical mereological principles.

Suppose x and y are superposed with one another, both of them occupying one and the same region of space R. It seems that the only compelling way to establish that some of the present properties of x are not among the present properties of y (i.e. that x is presently discernible from y) would be to indicate among the present properties instantiated in R a pair of *incompatible* present properties. If P_1 and P_2 are incompatible present properties, both instantiated in R, then P_1 and P_2 cannot both belong to x (or to y). Unless such a pair could be found, there would seem to be no reason to deny that all of the present properties instantiated in R belong both to x and to y.

If we had a developed theory of wholes as (concrete) distributive classes, rather than as sums, a plausible candidate for such a pair of incompatible properties might readily be found. We recall that the relation *is a member of*, characterising such classes, is intransitive - by contrast with the relation *is a part of* which characterises sums. Suppose it is proposed that x and y are distinct coinciding distributive classes. For example, x is the class of copper atoms in region R (the region occupied by the statue); y is the class whose members are two copper arms, two copper legs, a copper head and a copper torso, all found in R. On these assumptions, x has a relation to any of the copper atoms which is incompatible with the relation of y to that atom. If z is such an atom, then z is a member of x, but not a member of y.

Assuming, however, that both the lump of copper and the statue are sums, rather than distributive classes, no analogous incompatible (relational) properties of the superposed entities might be proposed. It appears, therefore, that although neoclassical assumptions countenance distinct superposed sums, they do not provide the resources to render such sums presently discernible from one another.

The assumption that distinct entities might be at the same place at the same time, or even superposed, at some time t, if it does not carry the further consequence that the entities are presently indiscernible, does not pose insuperable conceptual difficulties. Modern physics certainly urges us to accept that certain kinds of physical entities might be at the same place at the same time - for example, an electromagnetic field might be at the same place at the same time as a gravitational field.[23]

If we look over a wider range of categories, we find many cases in which we accept entities being in the same place at the same time without great difficulty: Two sounds, for example, may be simultaneously in one place; and on a trope-theory of universals, if a ball is red then there are distinct tropes which are in the same place (the region of space occupied by the ball) and at the same time. And as for superposition, it is not clear why it should be held to be more difficult to accept than being in the same place at the same time, at least if it is assumed that one of the sums is not identical to the entities of which it is a sum (or indeed if it is admitted that neither of them are identical to those entities).[24]

However, by contrast with the claims that distinct entities might be at the same place at the same time, or that they might be superposed with one another, the claim that they might be presently indiscernible at some time t does seem to present serious difficulties.

Suppose at t_1 one has identified a statue and a lump of copper which do not occupy exactly the same region of space at t. It is clear that they are not identical - if for no other reason than that one has parts which the other does not. Let us call the statue 's_1', and the lump 'l_1'. Suppose at t_2 one has again identified a statue and a lump of copper, but this time the identified entities do occupy exactly the same region of space and moreover they are superposed and indeed presently indiscernible at t_2. Let us call the statue identified at this time 's_2' and the corresponding lump 'l_2'.

Suppose that on some grounds (say, grounds of spatio-temporal continuity) one claims that the statue identified at t_2 is identical to the statue identified at t_1, and that the lump identified at t_2 is identical to the lump identified at t_1. Thus one is led to assume that at t_2 two distinct entities are presently indiscernible from one another, such that one of them is identical to s_1 and the other identical to l_1.

The difficulty that now arises is that there seems to be no non-circular answer to the question, what is it that makes it true that one of the coinciding entities rather than the other is identical to the statue that was identified at t_1 (i.e. to s_1). To see this, we first note that to say that - of the

---

[23] The sense of 'entity' required here is broader than the one usually found in this work.

[24] In Chapter 8 I shall argue that a *classical* mereological sum is identical to all its parts. This conclusion does not extend, however, to neoclassical sums.

two superposed entities s_2 and l_2 - it is the one that is identical to s_1 which is the one that is identical to s_1, is obviously circular. Secondly, to say that it is the one that is identical to s^* which is the one that is identical to s_1 (where s^* is a statue identified at some time other than t_1 and t_2, or at any time in some possible world other than the actual one) is again circular in that a similar question remains unanswered, namely, what is it that makes one of the superposed entities rather than the other identical to s^*.

Thirdly and finally, to say that of the two superposed entities it is the one that is a statue which is the one that is identical to s_1, leaves the problem equally unsolved. For given that the superposed entities are presently indiscernible, the property *is a statue* must be spelled out in terms of properties that are not present, and thus can be true only in virtue of an identity between one of the superposed entities and a statue identified at another time or in another possible world.

It seems then that we should accept the following as a principle governing identity statements:

Principle of Present Discernibility
PPD For all x and y, x is not identical to y if and only if
for all t, such that both x and y exist at t,
x is presently discernible from y at t

Assuming that the argument above is correct, if we wish to maintain (as I think we should) that distinct wholes may be superposed with one another, we require an account of wholes other than that according to which they are conceived as sums. For, as explained above, superposed sums must be presently indiscernible, and PPD requires superposed distinct wholes to be presently discernible. If we are to have superposed distinct wholes, we must have wholes that are not sums. The notion of non-unique neoclassical *sums* is incoherent.

III - Mereologically Varying Sums

To account for the *prima facie* fact that at least some types of wholes are susceptible to flux of parts, Peter Simons develops in some detail (as I have shown in section III of Chapter 5) several versions of the notion of a sum which is expressible using a temporally modified predicate. A sum of this sort is generally speaking an entity which is a sum-at-*t* of some entities, where to say that *y* is a sum-at-*t* of the *x*s is to say that for all *z*, *z* overlaps *y* at *t* if and only if *z* overlaps one of the *x*s at *t*. I will now show that a certain fundamental problem arises in connection with any attempt to account for the mereological variation of a whole in terms of such a temporally modified notion of a sum.

Let us say that if *y* is a sum-at-*t* of the *x*s, then *y* is a sum of the *x*s (at *t*). Therefore 'is a sum of ... at *t*' is defined as follows:

D1 *y* is a sum of the *x*s at $t =_{\text{def.}}$
for all *z*, *z* overlaps *y* at *t* if and only if *z* overlaps
one of the *x*s at *t*

The predicate 'is a sum of' is understood here as being temporally modified, whereas formerly in the course of the present work it was understood as a temporally unmodified predicate. To allow that it should designate either of the predicates should not lead to confusion, however. For it should be clear from the context whether one considers a certain entity to be a sum in the temporally modified sense of the predicate or in the temporally unmodified sense.

Consider a transition which a certain whole *y* undergoes between two particular moments of time, *t* and *t*'. Suppose that at *t* it is made up of the parts x_1 ... x_n and x^* (assume, for simplicity, that none of these parts overlaps any of the others at any time). Let us abbreviate the expression 'x_1 ... x_n' as 'the *x*s', and the expression 'x_1 ... x_n and x^*' as 'the *x*s+x^*'. The whole *y* is made up of the *x*s+x^* at *t*, and loses x^* (one of its parts) between *t* and *t*'. As a result it is made up of the *x*s at *t*'. To have a tangible example in mind, assume *y* is a house and that the *x*s+x^* are all bricks (this is to suppose, unrealistically, that the house is made up of bricks, without mortar, beams, etc.). x^* is a brick which is knocked out of one of the walls between four o'clock (*t*) and five o'clock (*t*') in the afternoon.

The whole *y* changes, then, from being made up of the *x*s+x^* at *t* to being a made up of the *x*s at *t*'. Accounting for the whole in terms of the notion of a sum, we assume that *y* is first a sum of the *x*s+x^*, and then changes into being a sum of the *x*s.

Two possibilities must be distinguished regarding the predicate 'is a sum of'. It is either a determinate predicate, or it is a vague predicate. If it is a determinate predicate, then at each moment of time *t*, and for all *u* and any *v*s, it is determinately the case or not the case that *u* is a sum of the *v*s at *t*. If the predicate is vague, on the other hand, there may well be a time *t* at which it is vague whether or not *u* is a sum of the *v*s at *t*. Consider first the assumption that 'is a sum of' is a determinate predicate.

1. Assuming 'is a sum of' is Determinate

Since in our example *y* changes from being a sum of the *x*s+*x** at *t* to being a sum of the *x*s at *t*', and since the predicate is determinate, it is clear that there is some point in time at which *y* ceases to be a sum of the *x*s+*x**. Let us call this moment 't_a'. Similarly, at some moment in time *y* begins to be a sum of the *x*s. Call that moment 't_b'. Note that we are assuming that both $t \leq t_a \leq t'$ and $t \leq t_b \leq t'$. We are not as yet making any further assumption regarding the relation between t_a and t_b (i.e. on our assumptions so far, t_a might be either before, at the same time as, or after t_b). In any case, however, let us call the period of time between t_a and t_b the 'transitional period'.

Let us now look more closely at the mechanism of the transition. It would seem that the transition must take place in one of the following three ways:

(1) The transitional period is of some finite length, and during this period *y* is both a sum of the *x*s+*x**, and a sum of the *x*s (i.e. $t_b < t_a$).

(2) The transitional period is finite, and during this period *y* is neither a sum of the *x*s+*x**, nor a sum of the *x*s (i.e. $t_a < t_b$).

(3) The transitional period is instantaneous (i.e. $t_a = t_b$).

None of these options seems acceptable, however. (1) is inconsistent with the definition of 'sum'. For on (1), there is some time *t* at which *y* is both a sum of the *x*s+*x** at *t*, and a sum of the *x*s at *t*. But if something is a sum of the *x*s at *t*, it cannot be a sum of the *x*s+*x** at *t* (assuming as we have that *x** overlaps none of the *x*s at any time). For, according to D1, *x** overlaps the former sum but not the latter one.

Option (2), while not leading to a contradiction, is nevertheless similarly untenable. It allows for two possibilities. Either

(2a) During the transitional period *y* is not a sum.[25]
or
(2b) During the transitional period *y* is a sum, but neither of the *xs*+*x**, nor of the *xs*.

According to (2a), if whenever *y* exists it is a sum, then *y* does not exist during this period. However, it is obviously implausible to assume that whenever an entity loses a part, there is a transitional period during which that entity does not exist. On the other hand, if *y* does exist during the transitional period, it cannot be a sum during this period. The conception of *y* as a sum is therefore inadequate, for being a sum is not an essential feature of *y*.

According to (2b), *y* is a sum of *some* entities at any moment during the transitional period, but neither of the *xs*+*x** nor of the *xs*. Now it would seem that at any time during the transitional period if *y* is a sum then (at least) all of the *xs* must be among the entities of which it is a sum. That is, *y* is a sum of the *zs*, such that the *xs* are all among the *zs* (in the example case of the house this means that all of the bricks that are not affected remain parts of the house throughout the transition). For both immediately before and immediately after the transitional period the *xs* are among the entities of which *y* is a sum.

Furthermore, on any plausible suggestion any entity from among the *zs* which is not among the *xs* must be a part of *x** (e.g. any part of the house during the transitional period, aside from the stable bricks, must be a part of the brick that is being knocked out). That is to say, a plausible reading of (2b) envisions a continuous process which takes place during the transitional period, where at t_a all the parts of *x** are parts of *y*, and gradually fewer of the parts of *x** are parts of *y*, until at t_b none of the parts of *x** are parts of *y*. When the brick is knocked out of the wall of the house, there is according to (2b) a transitional period during which the parts of the brick 'leak away' continuously, as it were, from being parts of the house to not being parts of the house, until at the end of the transitional period none of the parts of the brick are parts of the house.

This scenario is not plausible, however. For if we assume that during the transitional period the relations between the parts of *x** remain unchanged, there is no reason to think that some of the parts of *x** have

---

[25] I take the claim that *u* is a sum (at *t*) to be equivalent to the claim that for some *vs*, *u* is a sum of the *vs* (at *t*).

ceased to be parts of y, while others are still parts of y. It would seem that the parts of x^* stand or fall together, as regards their being parts of y. Any of the (indefinitely many) ways of specifying *which* parts of x^* have ceased to be parts of y at each point of time during the transitional period are bound to be arbitrary, or insufficiently motivated. For example, consider the difficulty in deciding whether the first parts of the brick which should cease to be parts of the house are concentrated at the centre of the far side of the brick, or are spread evenly over the whole of the far side of the brick. Moreover, no such proposal would be available in principle if there are *atoms* among the brick's parts. I conclude, therefore, that neither (2a) nor (2b) provide an acceptable account of the loss of a part., and so option (2) is disqualified.

Option (3) is similarly unacceptable. For it requires that the transition from x^*'s being a part of y to its not being a part of y take place *instantaneously*, in no time at all. Now, there can be no objection to the claim that the process whereby a part of an entity is lost may be *very brief*. The twig of a tree may cease to be a part of the tree with one fell swoop of an axe. Reflection on this case reveals however that we do not consider the separation of the twig from the tree to be strictly instantaneous. The course of the axe through the branch will require a fraction of a second.[26] If our reflection progresses down to the molecular level of the process, we still find only continuous processes, which take place in non-zero periods of time (e.g. the process of distancing of the last of the attached molecules of the twig).

Admittedly, according to a controversial account of vagueness (known as the 'epistemic view')[27] instantaneous changes should be acknowledged in connection with some types of qualitative change, such as the change from an apple's being green to an apple's being red. This account acknowledges that we are unable to determine a sharp transition between being green and being red, but maintains that this is neither to be attributed to a vagueness in our associated concepts, nor to a vagueness in the properties themselves, but rather to epistemological limitations. According to this view it seems necessary to assume that the transition from being green to being red takes place instantaneously (although we

[26] If, for example, the axe travels at 10 m/sec, and the twig is 1cm in diameter, then it would take 1/1000 sec. for the axe to traverse the width of the twig.

[27] This account of vagueness is extensively defended in Williamson 1994.

are unable, in principle, to determine the point at which this transition takes place).

However, the proposal that *macroscopic quantitative* changes (i.e. changes which are not infinitesimal, nor even microscopic in magnitude), as contrasted with *qualitative* changes, might take place instantaneously is fundamentally alien both to common sense and to physical theory (whether classical or modern). But the possibility of such changes is precisely what is required according to (3). If, for example, the brick weighs one kilogram and the house (before the brick is knocked out) weighs one ton, then the weight of the house is supposed to change instantaneously by one kilogram. And the volume of the house is supposed to change in a similarly abrupt way. If a tree is chopped down, and yet the stump continues to live, the tree will have instantaneously lost several tons in weight according to the present proposal.

Finally, even if it is suggested that a future physical theory might require us to accept the possibility of instantaneous quantitative changes, it seems highly undesirable that physical theory should be constrained to this result by mereological theory. It seems reasonable to require that mereology should be able to accommodate alternative possible courses of development in physical theory in this respect.

This concludes our investigation into the proposal that the loss of a part of a whole is to be understood in terms of the assumption that the whole is a sum, where the predicate 'is a sum of' is taken to be a *determinate* predicate. Let us now look at the alternative suggestion, according to which it is assumed that 'is a sum of' is *vague*.

2. Assuming 'is a sum of' is Vague

If 'is a sum of' is vague, and this vagueness is attributed to the *relation* designated by the predicate (thus rejecting a linguistic or epistemic account of vagueness in this connection),[28] then the following may be

[28] To assume the epistemic account would amount to repeating the proposal examined above; for on this account it is always determinately the case whether y is or is not a sum of the $xs+x^*$, only we are not always able to determine which is the case due to epistemic limitations. As for the linguistic account, it would seem to imply that the concept we associate with 'is a sum of' is not adequate for providing a *precise* characterisation of the relation between a whole and its parts; this seems to me to be a concession that the conception of wholes as sums is not ultimately satisfactory, which is what my criticism is meant to suggest.

suggested as an account of the transitional process whereby *y* loses one of its parts x^*.

Assume that the transitional period is of non-zero duration (thus avoiding the difficulties associated with instantaneous change of quantitative properties). Before t_a, *y* is (determinately) a sum of the *xs*+x^*. After t_b, *y* is (determinately) a sum of the *xs*. During the transitional period (between t_a and t_b) however there are no entities such that *y* is (determinately) a sum of those entities. For it is neither determinately true that *y* is a sum of the *xs*+x^*, nor determinately false that it is their sum; and similarly, it is neither determinately true that *y* is a sum of the *xs*, nor determinately false that it is their sum.

According to this account, then, on the occasion of the loss of a part, *y* changes from being determinately a sum of some entities (before the transitional period) to not being determinately a sum of any entities (during the transitional period), and then again to being determinately a sum of some entities (after the transitional period). This has none of the unwelcome consequences of alternatives (1), (2) and (3) above. No contradiction is arrived at as in (1), no implausibilites regarding the intermittent existence of *y* as in (2a), or the piecemeal process of losing x^* as in (2b), and no difficulties associated with instantaneous quantitative change as in (3).

However, it still seems that the notion of a sum understood in this way cannot be taken to be an adequate theoretical explication of the notion of a whole, in as far as the latter notion allows for mereological variation. For in taking *y* to be a whole we take it to be *determinately a whole* no more before and after the loss of its part x^* than during the transitional period in which that part is being lost. It is not the case that there is no fact of the matter, during the transitional period, whether or not *y* is a whole. The feature of being a whole is *determinately* preserved during circumstances in which a part is lost, much like the feature of being coloured is *determinately* preserved in circumstances in which the colour of an entity is changing.

On the other hand, we can see that in the transitional period there is indeed no fact of the matter, according to the present proposal, whether or not *y* is a sum. For it seems that for *y* to be determinately a sum at some time requires that for some entities *zs*, *y* is determinately a sum of those *zs* at that time. Affirming to the contrary would seem to be similar to claiming that *y* is determinately a lover at some moment of time without there being some *z* such that *y* is determinately a lover of *z* at that moment.

This difference points to a difference between the notion of a mereologically varying whole and a mereologically varying sum. The difference is particularly striking in the case of wholes which we consider to be continuously in a state of mereological flux, such as human organisms. An adequate account of the notion of a whole, in as much as we take this notion to apply to human organisms as paradigmatic cases, would be one which allowed us to say that under normal conditions the organism is determinately a whole. If we explicated the notion of a whole as the notion of a sum, we could not say this.

Indeed, if an alternative account were not available, we might have to concede that in as much as human organisms are mereologically varying entities, they cannot be determinately wholes. This might make Chisholm's mereological essentialism a more attractive position, for it denies that wholes - in the sense of 'whole' applying to *entia per se* (considered as genuine entities) rather than to *entia successiva* (considered as logical constructions) - are subject to mereological variation. On this view a human organism, far from being a paradigmatic whole, is rather a mere logical construction out of entities which cannot lose parts without ceasing to exist.[29]

To summarise, if the notion of a sum were indeed an explication of the notion of a whole, then we should expect vagueness to be found in the application of 'whole' in precisely those cases in which vagueness is found in the application of 'sum'. Since this appears not to be the case, we have substantial grounds for doubting that the former notion could be considered as providing an adequate account of the latter.

[29] See Chisholm 1976, 98.

Chapter 8
Sums, Collections and All the Parts

In the last chapter we saw that neoclassical mereology is only partly successful in accounting for features of wholes which were not satisfactorily accounted for by classical mereology. Numerous difficulties are associated with the ways in which neoclassical mereology accounts for important features of wholes - features whereby the existence of a whole seems to depend on the way its parts are conditioned; whereby more than one whole seems capable of being made up of precisely the same entities; and whereby wholes can survive the loss or gain of parts.

The question arises, what is the source of neoclassical mereology's weakness, that is, of its inability to break loose completely from the limitations which characterise classical mereology? A clue to an answer is pursued in section I. Neoclassical sums inherit a certain intrinsic formal limitation from their classical ancestors. I shall explicate this limitation, and indicate the way in which it renders the notion of a sum a version of the notion of a collection discussed in Chapter 2.

In section II I argue for a connection between the notion of a sum and the feature of non-monadicity. In particular, I argue that a classical sum is to be viewed as *identical* to its parts, and therefore a classical sum is not monadic.[1] This is a controversial point, and it is of interest independently of its role in my overall argument. The theory of Unities itself (which I present in Part Three) does not depend crucially on my claim that neoclassical sums are non-monadic, though to deny this claim would be to make that theory less elegant in a certain respect. However, to the extent that my conclusion regarding the non-monadicity of classical sums is acceptable, it helps (as we will see) to shed light from a different perspective on the distinction between sums and Unities.

---

[1] Monadicity was discussed above in Chapter 2, section II.

I - Inherent Limitation in the Notion of a Sum

It is clear from our preceding discussion that classical mereology proposes what may be described as a strict correspondence between entities and sums of those entities. Whenever there exist entities, the *x*s, there exists according to classical principles a unique entity *y* such that *y* is a sum of the *x*s. The transition from classical mereology to neoclassical mereology can be viewed as arising from an exploration of possibilities for loosening this strict correspondence. I wish to show now that the range of these possibilities is intrinsically limited by the way 'sum' is defined, and this limitation suggests that theoretical notions of a whole might be elaborated which do not share this limitation. The notion of a Unity will be seen to be precisely such a notion.

Neoclassical mereology denies the strict correspondence laid down by classical mereology according to which whenever there exist entities, the *x*s, there exists a unique entity *y* such that *y* is a sum of the *x*s. Instead, neoclassical mereology allows that for some *x*s there exists no entity *y* such that *y* is a sum of the *x*s. Of course, it also allows that for some *x*s, there exists a unique entity *y*, such that *y* is a sum of the *x*s. But it also allows, as a third option, that for some *x*s there exists a *non*-unique entity *y*, such that *y* is a sum of the *x*s. That is, in such a case there exists an entity *z*, such that $z \neq y$, where *z* is also a sum of the *x*s.

To summarise this contrast between neoclassical and classical mereology, let us first recall the definition of 'is a sum of':

<u>Definition of 'is a sum of'</u>
y is a sum of the *x*s $=_{\text{def.}}$
z overlaps *y* if and only if *z* overlaps one of the *x*s

The assumptions of classical and neoclassical mereology can be contrasted as follows:

Correspondence between entities and their sums - the classical position:
For any *x*s, for some *y*,
y is a sum of the *x*s, and for all *z*, if *z* is a sum of the *x*s then $z = y$.

Correspondence between entities and their sums - the neoclassical position:[2]
For any *x*s, either
1. for all *y*, it is not the case that *y* is a sum of the *x*s
or
2. for some *y*, *y* is a sum of the *x*s, and for all *z*, if *z* is a sum of the *x*s, then $z = y$
or
3. for some *y*, *y* is a sum of the *x*s, and for some *z*, *z* is a sum of the *x*s, and $z \neq y$

Notwithstanding this difference between the assumptions made by classical and neoclassical mereology, we find that there is an assumption which both theories share, an assumption which imposes a certain constraint as regards the correspondence between individuals and their sums. To see this, recall the notion of coextensiveness introduced in Chapter 3, section IV:

Definition of 'are coextensive with'
the *x*s are coextensive with the *y*s $=_{\text{def.}}$
for all *z*, *z* overlaps one of the *x*s if and only if *z* overlaps one of the *y*s

We find that both classical and neoclassical mereology make the assumption that if *u* is a sum of the *x*s and *v* is a sum of the *y*s, then $u = v$ only if the *x*s are coextensive with the *y*s. Using the terminology adopted in that section, this is to assume that sums are *coextensively determined* comprising entities. To demonstrate that both theories make this assumption is at the same time to explain why they are compelled to do so. For as we will see, this assumption follows from the very definition of 'is a sum of'.

[2] The claims I am making here apply both regarding the temporally unmodified predicate 'is a sum of' and the temporally modified one. If one wishes to be explicit about temporal modification, one would need to add quantification over time as well.

The proof proceeds by supposing that *u* is a sum of the *x*s and *v* is a sum of the *y*s, and that the *x*s are not coextensive with the *y*s, and showing that it follows from this together with the definitions of 'is a sum of' and 'are coextensive with' and with Leibniz's principle of indiscernibility of identicals that $u \neq v$.

Proof
(1) the *x*s are not coextensive with the *y*s [assumption]
(2) for some *z*, either
  (2a) *z* overlaps at least one of the *x*s but none of the *y*s
  or
  (2b) *z* overlaps at least one of the *y*s but none of the *x*s
  [from (1) and the definition of 'are coextensive with']
(3) (2a) [assumption]
(4) *u* is a sum of the *x*s, and *v* is a sum of the *y*s [assumption]
(5) *z* overlaps *u* [from (3), (4) and the definition of 'is a sum of']
(6) *z* does not overlap *v* [from (3), (4) and the definition of 'is a sum of']
(7) $u \neq v$ [from (5) and (6) and Leibniz's principle of indiscernibility of identicals]
(8) (2b) [assumption]
(9) $u \neq v$ [inference precisely analogous to that of (7)]

It follows therefore from (2), (4) and definitions and Leibniz's principle that $u \neq v$. Since (2) follows from (1) and the definitions, we have the required result: It follows from (1), (4), the definitions and Leibniz's principle that $u \neq v$. And so it follows from the definition of 'is a sum of' (granted the definition of 'coextensive' and Leibniz's principle) that if the *x*s and the *y*s have the same sum, they must be coextensive.

This constitutes a constraint on the type of correspondences which may be assumed to hold between entities and sums of those entities, a constraint with which any theory according to which wholes are conceived as sums must comply. This constraint expresses what might be described as a formal limitation inherent in the notion of a sum.

It is inherent in the definition of 'is a sum of', therefore, that sums are entities which are coextensively determined. Tracing the development of the notion of a sum from that of a collective class, we are assured that if the *x*s are concrete entities, a sum of the *x*s is a concrete, linguistically one, single entity. These features, together with that of being coextensively determined, guarantee that sums of concrete entities are collections, in the wide sense explained in section III of Chapter 3. As we shall see below, Unities are not collections in this sense. If *u* and *v* are Unities, and *u* comprises (i.e. is made up of) the *x*s, and *v* comprises the *y*s, then it may be the case that $u = v$ even in the case that the *x*s are not coextensive with the *y*s.

II - Classical Sums as Identical to their Parts

In the second section of Chapter 2 it was argued that if *y* is an entity which comprises the *x*s, the claims that the *x*s are many entities and that *y* is a single entity are compatible both with the *x*s' being identical to *y* and with the *x*s' not being identical to *y*. The supposition that the *x*s are many entities does not in itself compel us to think that the *x*s are not identical to *y*. This result allowed us then to assume that the comprised entities are many entities, and that the comprising entity is a single entity, while leaving open the question whether the comprising entity is identical to the comprised entities, and thus non-monadic.

Having examined the notions of a classical and a neoclassical mereological sum, it is now of interest to return to that open question. It is commonly assumed that *all* wholes, indeed, all comprising entities, are monadic, because the very idea of one-many identity is incoherent. This means that monadicity is a trivial feature of comprising entities. My purpose in this section is to de-trivialise monadicity, by arguing that classical sums are non-monadic. If this is correct, and *neoclassical* sums can be considered monadic only in virtue of *problematic* features associated with them (namely conditionality, non-uniqueness and mereological variation), features which I have criticised in the previous chapter, then *unproblematic* monadicity will be seen as a remarkable feature indeed, if anywhere to be found. And my eventual claim will be that Unities possess this remarkable feature. From this perspective, a connection will become visible between the explanatory advantages which I claim for the theory of Unities, and the fact that the notion of a Unity seems to offer a better grounded conception of a monadic comprising entity than does the notion of a neoclassical sum.

One terminological reminder before I proceed: In the discussion that follows I use expressions such as 'the parts', 'its part', and 'all the parts' interchangeably, to refer to entities, the *xs*, such that the whole is some *y* which is a sum of those *xs*. The context should make it clear (where clarity is desirable) precisely which of the whole's parts are those *xs*. By speaking of 'the parts' of a human organism, for example, one might mean molecules (where the organism is a sum of those molecules), or cells (where, again, the organism is a sum of those cells), or other parts. One does not usually (though one might) mean *both* the molecules *and* the cells. Context, as I've said, should determine which of these is intended, when something hangs on such possible differences.[3]

1. The Ambiguity

Let us first note that any claim of the form

(1) *y* is identical to the *xs*.

(or equivalently 'the *xs* are identical to *y*') is ambiguous, in that it can be understood in one of two ways: *distributively* or *collectively*. Taken distributively it is synonymous with

(1a) *y* is identical to *each* of the *xs*.

Taken collectively, it is synonymous with

(1b) *y* is identical to *all* of the *xs* *taken together*.

The phrase 'taken together', in the latter sentence, is of course metaphorical. To explain its significance, and the significance of the contrast between the two readings of '*y* is identical to the *xs*', we may consider an analogous contrast between different readings of sentences in which the predicate is not 'is identical to' (or 'are identical to') but rather some predicate of the form 'is identical in respect of *A*'. Consider, for example, the sentence

(2) The people are identical in (i.e. in respect of) weight to the stone

[3] See in this connection my remarks on 'the parts' in Chapter 4, section I.

This sentence can be read in two ways. Read in one way, it is synonymous with

(2a) Each of the people is identical in weight to the stone

Read in another way, it is synonymous with

(2b) All the people taken together are identical in weight to the stone

The latter sentence (2b) is understood as implying that the (arithmetical) sum of the people's weights is identical to the weight of the stone. The former sentence (2a) clearly does not imply this, for in case the number of people is greater than one (2a) is not even compatible with this implication. Therefore it is clear that the two readings are not equivalent to one another.

To claim that '*y* is identical to the *xs*' can be taken either distributively or collectively is to claim that this sentence is susceptible to alternative readings which contrast in a way analogous to the two readings of the sentence 'the people are identical in weight to the stone'. The analogy also leads us to assume that the two readings of the former sentence, (1a) and (1b), are not equivalent to one another. Let us say that the statement '*y* is identical to the *xs*', taken distributively, is a *distributive identity* statement, while the statement, taken collectively, is a *collective identity* statement. To distinguish the two readings I will use '*y* is distributively identical to the *xs*' when distributive identity is meant, and '*y* is collectively identical to the *xs*' when collective identity is meant.

It would seem wrong to interpret contrast between distributive and collective identity statements in terms of a distinction between *kinds* of identity, or *senses* of 'identity'. Compare this with the claim that Samson lifted two stones, which might be understood as the claim that he lifted the stones either at different times, or at the same time. It would be strange to assume that this is to distinguish between two kinds of lifting, or two senses of 'lift'.

It would be more plausible to suggest that the distinction between the two types of statement hinges on the question of which entities are asserted to stand in the relation *is identical to* to one another. There is a unique relation *is identical to*, but in a distributive identity statement *y* is asserted to stand in this relation to each of the *xs*, while in a collective identity statement this is not asserted, but rather that *y* stands in this relation to all of them collectively.

Paralleling the two interpretations of (1), we have two corresponding interpretations of sentences of the following form:

(3) *y* is distinct from the *x*s.

Namely,

(3a) *y* is distinct from each of the *x*s.

(3b) *y* is distinct from all the *x*s taken together.

In the case of (3a) we will say that *y* is *distributively distinct* from the *x*s, while in the case of (3b) we will say that *y* is *collectively distinct* from the *x*s. Note that whereas (1b) and (3b) are contradictories, (1a) and (3a) are merely contraries.

2. Objection on Categorial Grounds

Commenting on what Doepke calls 'The One-Many View', Simons notes that this view 'disallows (quite properly) identity between a plurality of objects, such as a number of wooden boards, and a single object, such as a ship, made up of them'.[4] Both Doepke and Simons deny that this view provides good reasons for rejecting Wiggins' conclusion that distinct entities may be superposed with one another. However, both seem to think that the principle on which the response of upholders of the One-Many View is based is true. Thus van Inwagen, Doepke, Simons, and anyone who subscribes to the One-Many View, are apparently all in agreement that statements of the form (1) should be 'disallowed', unless the *x*s are all identical to one another.[5]

Now, it is clear that (1a) can only be true if the *x*s are all identical to one another. Denying this would lead immediately to a contradiction. For suppose a and b are among the *x*s, such that $a \neq b$. From (1a) it follows that $y = a$ and that $y = b$. Assuming that identity is symmetric and transitive, this implies that $a = b$, contradicting the assumption that $a \neq b$.

---

[4] Simons 1987, 212.

[5] See van Inwagen 1990, 287 (note 13); Doepke 1982, 46-7; Simons 1987, 212.

However, it seems that no contradiction can be derived from (1b). Presumably the reason for disallowing (1b) in case the *x*s are not all identical to one another would then be that this statement involves some sort of category mistake - that it is in some way nonsensical.

It is not clear however why (1b) should be thought non-sensical in such non-trivial cases (i.e. cases in which the *x*s are not all identical to one another). Perhaps the reason for thinking this is that it is commonly supposed that *is identical to* is a two-place relation.[6] Certainly this is supposed with regard to the relation as it is understood in the context of standard formal logic (i.e. predicate calculus with identity). If *is identical to* is a two-place relation then to claim, for example, that distinct entities *x* and *y* are collectively identical to an entity *z* is to fail to understand the predicate 'is identical to', taking it to designate a three-place relation, or a multigrade relation, instead of a two-place one. A mistake would be involved here which is similar to the mistake involved in claiming that *x* asks *y*, while denying that there is some *z* such that *z* is what *x* asks *y*. In this latter example, the mistake would consist in taking a triadic predicate ('asks') to be a dyadic predicate.

The fact that *is identical to* is understood as a two-place relation in the context of standard logic does not, however, provide sufficient evidence for the view that statements of the form (1b), in non-trivial cases, are nonsensical. For it is widely recognised that standard logic is incapable of representing the full range of valid inferences made in common discourse. And the question of how standard logic must be supplemented in order to represent adequately the inferences of common discourse is still far from settled.[7] Just as it is appropriate to inquire whether the reduction of plural quantification to singular quantification is generally justified, so it would seem appropriate to inquire whether the reduction of collective identity to distributive identity is generally justified.

On the positive side, Donald Baxter has helpfully pointed out that common sense judgements often seem to betray the assumption that statements of the form (1b) in many cases not only make sense, but are true. As he plausibly remarks:

Someone with a six-pack of orange juice may reflect on how many items he has when entering a 'six items or less' line in a grocery store.

---

[6] See, for example, Russell 1903, 95-6; Grossmann 1983, 170-2.

[7] On this question, see Boolos 1984.

He may think he has one item, or six, but he would be astonished if the cashier said 'Go to the next line please, you have seven items'. We ordinarily do not think of a six-pack as seven items, six parts plus one whole.[8]

It is a common sense view that in this case we have six bottles of orange juice, and that at the same time we have one six-pack of orange juice, and yet we do not have seven items. Rather, we have either six items or a single item, depending (one might suggest) on the way we wish to describe what we have. The various judgements involved here are compatible if and only if the six items are identical to the single item. *Prima facie*, then, it would seem that in some cases it is true to claim that y is collectively identical to the xs.

3. Arguments Purporting to Refute the Thesis

Assuming then that statements of the form (1b) are acceptable at least from the categorial perspective (that is, assuming that they make sense), there seems *prima facie* to be a very strong case indeed for claiming that a classical sum is collectively identical to its parts. First of all, this is suggested by the fact that the sum and the parts (collectively) have precisely the same identity conditions. If y is the (classical) sum of the xs then y exists if and only if the xs exist.

Secondly, it is suggested by the fact that the sum occupies at all times precisely the same region of space as the parts (collectively) do. Indeed, if a single entity x (rather than many entities, the xs) stood in such relations to y, there would hardly be any doubt in our minds that y is identical to x. It seems to me that these considerations clearly place the burden of proof on those who would argue that the sum is *not* identical to its parts.

Therefore, I propose in what follows to consider several arguments which purport to show that a sum is not identical to its parts. To the extent that the answers that I offer in response to these arguments are satisfactory, I shall consider the claim that the sum is identical to its parts to have been established, pending the presentation of new reasons for denying this claim.

I shall begin by discussing an argument which Donald Baxter considers to be the principal argument against the view that a sum is

[8] Baxter 1988, 579.

identical to its parts.[9] Baxter contrasts two views that one might take regarding sums, the Identity view (which I call 'SIP') and the Non-Identity view (which I call 'SDP'):

SIP The sum is identical to all its parts

SDP The sum is distinct from all its parts

Position SIP, he claims, is the one commonly taken by people who are not professional philosophers. By contrast, in professional philosophical contexts the more commonly held position is the SDP.[10] Baxter thinks that this divergence between common sense and professional opinion is to be explained by the fact that philosophers are confronted with the following argument (which I call 'Argument I') against SIP, which is 'straightforward and compelling':[11]

Argument I

The sum and all the parts exist. The sum comprises all the parts. None of the parts do. So the sum has a (relational) property none of the parts has. So the sum is identical with none of the parts. So it is distinct from each. So if n parts exist and the sum exists, then (at least) $n+1$ things exist.[12]

---

[9] See Baxter 1988. Throughout his discussion, Baxter uses the term 'whole', rather than 'sum' or 'classical sum'. As I have noted above, however, there are straightforward reasons for thinking that a neoclassical sum is *not* identical to its parts. The reasons he gives, both for and against the view that the whole is identical to its parts, are at their most plausible if phrased with respect to the particular type of wholes we have been calling 'classical sums'. To keep this clearly in mind, I have substituted his 'whole' for 'sum' throughout, where I take 'sum' as short for 'classical sum'.

[10] Baxter 1988, 578. Of course, people who are not professional philosophers are more likely to use the term 'whole' than 'sum' (indeed, they are not likely to debate about classical mereological sums). However, Baxter's point is that in common sense thinking about wholes *in general* there is an inclination to assume that the whole is identical to its parts. Thus if common sense intuition is to be represented in a debate about classical sums, this would reasonably be done by attributing to common sense (according to Baxter) the belief that a sum is identical to its parts (SDP).

[11] Baxter 1988, 578.

[12] Baxter 1988, 578-9, substituting 'sum' for 'whole'.

Prior to the last sentence, this argument is concerned with establishing that the sum is distinct from *each* of its parts. Argument I consists, then, of the following three steps

(i) The sum is distinct from each of the parts

(ii) If n parts exist and the sum exists, then (at least) $n+1$ things exist

Therefore,

(iii) The sum is collectively distinct from its parts

Baxter first considers a response to this argument which he finds unsatisfactory.[13] He then presents his own response, according to which it is denied that (ii) follows from (i). And since it is presumably clear that (iii) does not follow from (i) on its own, this blocks the inference to (iii).

---

[13] This response expresses what he calls the 'Combination' view, according to which

... if there are six things then a seventh thing - the sum they are parts of - exist, and the six things collectively are the seventh thing. (Baxter 1988, 579; as before, substituting 'sum' for 'whole'.)

He explains that the Combination view involves the assumption that a sum is a *multitude*, rather than a *single thing*, where it is assumed, nevertheless that a multitude is *one thing* no less than a single thing is. Baxter apparently takes it that the sum's being one thing would explain why it counts as a *seventh* thing, while its being a multitude (rather than a single thing) would explain why this is compatible with its being collectively identical to the parts.

I have to admit that I find this response obscure. In the first place, I don't understand his contrast between a multitude and a single thing. The multitude is asserted to be one thing, and yet I cannot make out a sense of 'being one' which would suit the argument's purposes. In the second place, even if a consistent sense could be given to the terms he uses, the argument does nothing in the way of explaining *how* it is that if an entity counts as a seventh thing, that is as something in *addition* to six things, it may nevertheless be (collectively) identical to the six things. This seems to ignore the widely acknowledged close conceptual connection between counting and identity.

However, as I have noted, Baxter himself finds this response unsatisfactory, although for different reasons. His main reason seems to be that the Combination view entails that all sums are multitudes, and this, he claims, commits us to atomism, and "one's theory of parts and whole should not mandate atomism". However, his response suffers from the unclarities noted with regard to the notion of a multitude, no less than does the Combination view itself. See Baxter 1988, 580.

This proposal involves denying that the sum is something in addition to its parts (i.e. denying (ii)), as a means for upholding the claim that the sum is collectively identical to the parts. He thinks that such a proposal can be justified by appealing to Butler's view of Identity:

> Assume that on strict standards for counting the parts are many and on loose standards they are one. The strictly distinct parts are identical with each other on a loose standard. The whole, then, is just the parts counted loosely. It is strictly a multitude and loosely a single thing.[14]

The example by means of which Baxter illustrates the distinction between counting on strict standards and counting on loose standards involves a sapling and the mature oak tree into which it develops with the passage of time. The sapling and the mature oak are taken to be strictly distinct temporal parts of the temporally extended oak tree.

> Counting according to strict standards (according to Butler) the sapling is one thing and the mature tree is another. Counting according to loose standards the sapling is one thing and the mature tree is that one thing again.[15]

Baxter's treatment of Argument I requires us to apply a similar distinction in connection with counting the (contemporaneous) parts of a sum. Suppose x and y are what we normally take to be distinct parts of a sum (say, two of the orange juice bottles, which are distinct parts of the sum of six such bottles). Then, counting on strict standards, x is one thing and y is another. Counting according to loose standards, x is one thing and y is that one thing again.

The resulting picture is as follows: Counting strictly there are n entities. Counting loosely, there is one entity. Under neither standard of counting do we arrive at $n + 1$ entities. Thus assumption (ii) is denied.

I find this solution unconvincing. The notion of a loose identity as required by Baxter is a strange one. For on his account, entities x and y are loosely identical to one another if and only if there is some entity z of which they are both parts (note the generality implied in the quote under 'Argument I' above). Thus, for example, if the Earth is a sum, and Mount Everest and Loch Lomond are parts of the Earth, then Mount Everest and Loch Lomond are supposed to be loosely identical to one another. But

---

[14] Baxter 1988, 580. With regard to Baxter's notion of a *multitude*, see previous footnote.

[15] Baxter 1988, 576-7.

clearly the relations between them bear little if any analogy to the relation *is identical to*, so that the use of the predicate 'is loosely identical to' would not be appropriate with regard to them, unless by virtue of a completely altered meaning of 'identical' as featuring in that predicate. But if Baxter uses 'identical' in a radically altered sense, the assumptions which his argument requires - connecting the notion of identity with the notion of counting (for he says that since the parts are loosely identical, they count on loose standards as *one* entity) - seem to be unfounded.

Indeed, one would be forgiven for thinking that what Baxter means by 'is loosely identical to' is simply 'is a part of the same sum as'. But in itself, the fact that the relation *is part of the same sum as* holds between the parts of a sum gives us no reason to think that the sum is collectively identical to its parts.

I would suggest that Baxter's account is intuitively grounded in the undeniable observation that there is a certain sense in which we might speak of many entities (even if radically distinct from one another) 'as one'. This, however, should in no way incline us to suppose that in such cases we treat the many entities as 'loosely identical to one another'. Rather, I think that the correct account of what we mean in such cases is as follows: Instead of simply speaking of the many entities, we speak in such cases of a single entity that purportedly comprises those entities. To describe speaking of the comprising entity as a way of speaking about the many comprised entities is perhaps acceptable even in cases where the comprising entity is deemed not to be identical to the comprised entities, though regarding such a case the description would indeed be loose. But as I said, the looseness here does not concern a supposed identity of the comprised entities to one another.

Notwithstanding my rejection of Baxter's response, however, I think that a plausible account is available by reference to which the inference of Argument I can be blocked.

Argument I is based on certain presuppositions concerning the relation between judgements as regards number (i.e. counting) and judgements as regards distinctness. Now, while it could hardly be denied that there is some such relation between the two types of judgement, I will show that Argument I is based on an inconsistent account of this relation.

Consider the second step of Argument I. Inspection shows that the derivation of (ii) from (i) presupposes the following principle relating distinctness and counting:

A if the *xs* are *n* in number then
if *y* is distributively distinct from the *xs*, then the *xs*+*y* are $n+1$ in number

Principle A is plausible. An example: If those bottles are 5 in number then, if this bottle is distributively distinct from those bottles (i.e. distinct from each of those bottles), then those bottles and this are 6 in number. Another example: If those bottles are 6 in number then - if this pack (i.e. a six-pack of bottles) is distributively distinct from those bottles, then the bottles and the pack are 7 items in number.

The converse of A is equally plausible. It would make little sense to assume that those bottles are 5 in number, and those bottles and this are 6 in number, and yet this bottle is not distributively distinct from those. Therefore, A is reasonably viewed as a partial statement of the following more complete principle:

A* if the *xs* are *n* in number then
y is distributively distinct from the *xs* if and only if the *xs*+*y* are
$n+1$
in number

Next, consider the second step of Argument I. Again inspection reveals that the derivation of (iii) from (ii) presupposes the following principle relating distinctness and counting:

B if the *xs* are *n* in number then
if the *xs*+*y* are $n+1$ in number then *y* is collectively distinct
from the *xs*

Principle B is also plausible. If those bottles are 5 in number then - if those bottles and this one are 6 in number, it follows that this bottle is collectively distinct from those bottles (i.e., this bottle is distinct from those bottles taken together). Similarly, if the bottles are 6 in number then - if the pack and the bottles are 7 items in number then the pack is collectively distinct from the bottles (i.e., again, the pack is distinct not only from *each* of the bottles, but is also distinct from all of them taken together).

The converse of B, however, is not plausible - indeed, it is clearly false. According to the converse of B, supposing b_1 , b_2 and b_3 are three bottles, and bottle b_3 is collectively distinct from b_1, b_2 and b_3 (for b_1 is distinct from b_1, b_2 and b_3 taken together), it follows that b_3 and b_1, b_2 and b_3 are 4 in number. But this is false.

This suggests that a more complete principle is presupposed in the derivation of (iii) from (ii), a principle that implies B, and whose converse is as plausible:

B' if the *xs* are *n* in number then
if the *xs*+*y* are *n*+1 in number then *y* is collectively
and distributively distinct from the *xs*

Both B' and its converse seem unobjectionable, as experimenting with examples of bottles and packs quickly shows. Combining the B' with its converse, we arrive at the following principle:

B* if the *xs* are *n* in number then
y is collectively *and* distributively distinct from the *xs*
if and only if the *xs*+*y* are *n*+1 in number

Comparing A* and B* we see that Argument I makes use of two distinct principles regarding the connection between distinctness and counting. Significantly, to assume that the two principles are compatible with one another is to assume that if *y* is distributively distinct from the *xs* then it is also collectively distinct from the *xs*. But this is precisely what Argument I sets out to establish. So that in making use of *both* principles, Argument I is either simply begging the question, or guilty of inconsistency. For on inspection it is clear that if one consistently used either of these principles in Argument I, step (iii) could not be derived. If (A*) were used, then step (iii) could not be derived from (ii). If (B*) were used, then step (ii) could not be derived from (I). We may conclude then that Argument I must be rejected.

As to the question, which, indeed, of the two principles A* and B* is the correct one, I think this is most plausibly viewed as a matter of convention. In the context of ordinary everyday counting, both yield the same results. Suppose we are counting the oranges in a plastic bag, pointing to each of them consecutively. Suppose we have already counted five of them. Pointing now to an orange, we may well be in doubt as to whether it is identical to one of those already counted. That is, we may well be in doubt about whether this orange is distributively distinct from the oranges already counted. We are always clear, however, that the orange is collectively distinct from the oranges already counted.

In the more unusual cases where the different principles yield different results, when one is asked how many entities there are here, it is appropriate to reply: 'How many on which counting convention?'. Returning to the case of the six bottles and the six-pack of those bottles, according to A* one would say that there are seven (whether or not the sum is collectively identical to its parts); according to B* one would say either that there are six (assuming that the sum is collectively identical to its parts), or that there are seven (assuming that the sum is collectively distinct from its parts).

David Lewis supports a view which comes very close to affirming SIP, and takes his own view to be sympathetic, and similar, to that of Baxter.[16] He calls his position on this matter the thesis of Composition as Identity. He too, however, is not convinced by Baxter's argument, and as a result makes his position out to be somewhat weaker than an unqualified embracing of SIP (and so weaker than the position I am arguing for). He claims that mereological relations, such as *compose*, *is a part of*, and *overlaps*, are

> strikingly analogous to ordinary identity, the one-one relation that each thing bears to itself and to nothing else. So striking is this analogy that it is appropriate to mark it by speaking of mereological relations ... as kinds of identity. Ordinary identity is the special, limiting case of identity in the broadened case.[17]

With regard to the relation between a single part and its sum (assuming that the sum has parts which are not identical to that part), the claim for an analogy with identity is related to David Armstrong's position, according to which Australia and New South Wales, for example,

> are not identical, but they are not completely distinct from each other. They are partially identical.[18]

[16] Lewis 1991, 81-7.
[17] Lewis 1991, 84-5.
[18] Armstrong 1978, Vol. II, 37-8.

It is of course clear that the relation between a single (proper) part and its sum can at most be taken to be analogous to identity.

Regarding the relation between *all* the parts and their sum, Lewis interestingly points out several ways in which it is analogous to identity, ways which he regards as very strong evidence for the conceptual proximity between composition and identity.[19]

Why then does Lewis refrain from simply embracing SIP? it is important for him after all to assert that (classical) mereology is 'ontologically innocent', that is, that if one starts by commitment to the existence of several individuals, and adopts the assumption that a sum of those individuals exists, one does not thereby increase one's original commitment. And this assertion would be made more compelling if it could be claimed that a mereological sum is identical to all of its parts (rather than merely related to all its parts in a way analogous to identity).[20]

He presents two arguments for rejecting this otherwise desirable view. Neither of them, however, seems to me to be convincing. I shall now examine them in turn, dubbing them Argument II and Argument III, respectively.

Argument II

> We know that *x* and *y* are identical if and only if, whenever there are some things, *x* is one of them if and only if *y* is one of them. But if *y* is the sum of the *x*s, then there are some things such that each of the *x*s is one of them and *y* is not; and there are some things such that *y* is one of them but none of the *x*s is.[21]

This argument seems to be confused. Indeed, we may accept that *x* and *y* are identical if and only if whenever there are some things, *x* is one of them if and only if *y* is one of them. And we may also accept that if *y* is the sum of the *x*s, then there are some things such that each of the *x*s is one of them and *y* is not. For example (assuming that the *x*s are all *proper*

---

[19] See Lewis 1991, 84-87.

[20] Simons takes Lewis's ultimate rejection of the claim that the sum is identical - strictly speaking - to its parts, to be incompatible with Lewis's earlier claim that 'given a prior commitment to cats, say, a commitment to cat-fusions is not a *further* commitment'(Lewis 1991, 81). See Simons 1991, 396.

[21] Lewis 1991, 87; substituting 'sum' for 'fusion'.

parts of y), each of the xs is one of the proper parts of y, while y is not one of the proper parts of y.

This, however, does not seem relevant to the denial of SIP. For it only shows that *each* of the xs is not identical to y. That is, it shows that y is distributively distinct from all the xs. It gives us no reason to think that y is collectively distinct from all the xs. SIP, we recall, is the claim that the sum is collectively identical to all its parts.

Similarly, we may accept that there are some things such that y is one of them but none of the xs is. For example, y is one of the things of which all of the xs are parts, while none of the xs is one of those things (again, assuming that the xs are all proper parts of y). Again, this only shows that y is distributively distinct from all the xs.

Argument III

> ... even though the many and the one are the same portion of Reality, and the character of that portion is given once and for all whether we take it as many or take it as one, still we do not really have a generalized principle of indiscernibility of identicals. It does matter how you slice it - not to the character of what's described, of course, but to the form of the description. What's true of the many is not exactly what's true of the one. After all they are many while it is one. The number of the many is six, as it might be, whereas the number of the sum is one. And the singletons of the many parts are wholly distinct from the singleton of the one sum.[22]

Lewis is claiming here that whenever y is a sum of the xs, certain predicates apply to y which do not apply to the xs, and conversely certain predicates apply to the xs which do not apply to y. If this is the case, then indeed to claim that y is collectively identical to the xs would be to violate the principle of indiscernibility of identicals, and therefore this latter claim should be rejected.

Let us consider his position more carefully. Let our example case be that of six horses and their sum. Lewis maintains that these differ because (1) the horses are many while the sum is one; (2) the number of the horses is six while the number of the sum is one; and (3) the singletons of the

[22] Lewis 1991, 87; again, substituting 'sum' for 'fusion'.

horses are 'wholly distinct' from the singleton of the one sum. Point (3) is very different from the first two points, and may be dealt with straightforwardly.

It seems that in the case of (3) we have again a confusion of the sort that was noted in Argument II. Granted, u and v cannot be identical unless they have the same singletons. And granted, the sum of horses has a singleton which is distinct from the singletons of each of the horses. This only shows, however, that the sum is distinct from each of the six horses, that it is *distributively* distinct from the horses. This is hardly a difficult claim to accept, on any plausible view. But as I have already pointed out, distributive distinctness does not imply collective distinctness. On the contrary, had the sum of horses not been distributively distinct from the six horses, it would be impossible for it to be collectively identical to them. Lewis's point (3), then, gives us no reason at all to doubt that the horses can be collectively identical to the sum.

Points (1) and (2) do not involve a confusion of this sort. It is not completely clear what Lewis thinks the conflict is, exactly. Presumably regarding (1) he takes it that the six horses are many while the sum *is not* many; and that the sum is one, whereas the horses are not one. It is by no means clear, however, why the sum is supposed not to be many, and why the six horses are supposed not to be one. To make such claims would not seem to put too great a strain on linguistic usage. Indeed, what is known as the 'constitutive' use of the copula involves precisely such statements, at least when oblique statements are used: 'the many molecules are (i.e. constitute/compose/make up) one table' and 'the table is (i.e. is constituted by /composed of/made up of) many molecules'; and hence, obliquely, 'the many molecules are one' and 'the table is many'.

Perhaps Lewis takes (1) to be simply making the same point as (2), only formulated somewhat differently. Now, in connection with (2) it *does* seem that a sentence like 'the number of the sum is six' would be seriously at odds with accepted usage, by contrast with 'the number of the horses is six'. This, however, is still far from showing that the horses are not collectively identical to the sum. Consider the analogous question, whether it is possible for a man to be identical to a woman. Surely this question cannot be settled by the merely linguistic consideration that a sentence such as 'he talked to herself' is at odds with usage (perhaps the question is conceptual; perhaps empirical).

Perhaps the most natural interpretation of the points Lewis is making in (1) and (2) is not one that takes them to rely on linguistic considerations. According to this interpretation, the claim that the horses are many, for example, means the same as the claim that the horses are (i.e. are identical to) many horses. And the claim that the sum is one means the same as the claim that the sum is (i.e. is identical to) one sum. Similarly, the claim that the number of the many is six means the same as the claim that the many are six horses; and the claim that the number of the sum is one means the same as the claim that the sum is one sum.

If this is how Lewis's claims are to be taken, however, it is not clear on what grounds he is denying that what is true of the many is also true of their sum. If the horses are (i.e. are identical to) many horses, it is not clear on what basis it is being urged that the sum of horses is not also (i.e. is not also identical to) many horses. And if the sum is (i.e. is identical to) one sum, it is not clear on what basis it is being denied that the horses are (i.e. are identical to) one sum.

The claims in the last paragraph do not seem to involve linguistic impropriety (see my comments several paragraphs above concerning the constitutive use of the copula). To reject them as false because of the principle that many entities cannot be identical to one, or in particular that many parts cannot be collectively identical to their sum, is of course to beg the question. For Lewis's argument attempts to lead precisely to that principle as its conclusion.

Therefore, as far as I can see, Lewis's discussion gives us no substantial reasons for thinking that a sum cannot be collectively identical to its parts. All in all, the evidence we have found for the assumption that a sum is collectively identical to its parts well outweighs the evidence against it.

Part III: Unities

Chapter 9
A Theory of Unities

I - Introduction

In section I of Chapter 8 we observed that a certain constraint follows from the way in which the notion of a sum is defined, with regard to the types of correspondences allowed between individuals and their sums. Thus we found that if the *x*s and the *y*s have the same sum (at *t*), they must be coextensive (at *t*). This constraint provides us with a clue as to a way in which an account of wholes might be given according to which they are not construed as sums. To see this, however, we need to take a step back and examine the notion of a whole with a view to explaining, rather than simply taking for granted, its alleged connection with the notion of a sum.

Before embarking on this in the next section, however, I would like to give an overview of the idea I shall be pursuing, for fear of the many trees blocking the view of the forest later, when we enter various details. I shall describe therefore the picture in broad outline, postponing the detailed justification for the suggestions until such a first sketch is already before us.

We begin by noting again that when we say that certain parts *make up* a whole, we mean more than that each of them is a part of the whole. Indeed, the claim that they make up the whole entails that each of them is a part of the whole. However, the claim that each of them is a part of the whole does not entail that they make up the whole. My fingers are parts of my body, but they do not make up my body.

This indicates that between wholes and their parts two relations must be distinguished. One is the relation which a whole has to its part *distributively*, that is, a relation which consists in *each* of the parts bearing the relation *is a part of* to the whole. The other is the relation which a whole has to it parts *collectively*, a relation consisting in the parts - considered *together* - bearing the relation *make up* to the whole. I distinguish therefore between two 'whole-parts' relations: The distributive whole-parts relation, which I call 'is a whole with respect to'; and the collective whole-parts relation which I call 'is a whole which corresponds to'.

The conception of wholes as sums is founded on the assumption that the collective whole-parts relation is none other than the relation *is a sum of*. This involves, as we have seen, the assumption that wholes are coextensively determined. I describe this as the view that wholes are *fundamentally* sums. According to this view therefore *y* is a whole which corresponds to the *x*s only if *y* is a sum of those *x*s. It is remarkable, however, that this assumption is not a consequence of the principles adopted by neoclassical mereology (nor even those of classical mereology). One may assume that the relation *is a proper part of* is irreflexive and transitive, and that it complies with the principle of Supplementation of Proper Parts, and even with the classical assumption connecting *is a part of* with *is a proper part of* and *is identical to* - and yet consistently assume that something is a whole which corresponds to the *x*s without being a sum of those *x*s.

A theory may be proposed, therefore, which accepts the latter fundamental mereological assumptions, and yet rejects the assumption that the collective whole-parts relation is necessarily the relation *is a sum of*. Indeed, it rejects the assumption that all wholes are coextensively determined. According to this theory, a Unity is an entity which may well be a whole which corresponds to the *x*s without being a sum of the *x*s. A Unity therefore is a whole which is *not* fundamentally a sum. It is still the case that a Unity is a sum of *some x*s, however. This is what I mean by saying that it is *peripherally* a sum. It is a whole which corresponds to those *x*s and at the same time to some *y*s, where it is a sum of the *x*s but not a sum of the *y*s.

Suppose for example that a tree has two main branches, Branch1 and Branch2. Let us assume then that it is made up (at some time *t*) of Root, Trunk, Branch1 and Branch 2. According to the conception of wholes as sums, this implies that the tree is *not* made up (at *t*) of Root, Trunk and Branch1. Anything that is a sum of the latter cannot be identical to something that is the sum of Root, Trunk, Branch1 *and* Branch2. For the parts which exclude Branch2 are not coextensive with the latter parts, which include Branch2.

According to the conception of wholes as Unities by contrast it may well be the case both that Root, Trunk, Branch1 and Branch2 make up the tree (at *t*), and that Root, Trunk and Branch 1 make up the tree (at *t*). Unities are not assumed to be coextensively determined, and a Unity may be a whole which corresponds to certain parts without being a sum of those parts. Indeed, to assume that the tree is a Unity is not to deny that it is *peripherally* a sum. The tree, in this case, is clearly a sum of Root, Trunk, Branch1 and Branch2. However, the tree is not fundamentally a sum, and so it can be made up also by parts of which it is not a sum.

I shall be assuming that whenever the *x*s make up a Unity, and each of the *x*s is an *immediate* proper part of the Unity (i.e. it is not a proper part of any proper part of the Unity), then the *x*s *underlie* the Unity. The relation *underlie*, then, is the relation which immediate proper parts - taken collectively - have to the whole, if that whole corresponds to them. An immediate proper part of a Unity is called an *element* of that Unity. A proper part of a Unity, in general (i.e. whether immediate or not), is called a *pre-element* of the Unity. The version of the theory of Unities which I propose here is developed in terms of assumptions about the relations *underlie*, *is an element of*, and *is a pre-element of*.

A more comprehensive account would treat not only *underlie*, but also other types of the relation *make up* which apply in connection with Unities - that is, types which apply not only between a Unity and its *immediate* proper parts. In the present context I shall confine myself to the discussion of *underlie*, however, without pretending to treat the theory of Unities exhaustively and conclusively.

Example: If Root, Trunk, Branch1 and Branch2 are assumed to be *immediate* proper parts of the tree, then they will be said to be elements of the tree, and (taken collectively) to underlie it. To assume that these elements underlie the tree is consistent with assuming that Root, Trunk and Branch1 (without Branch2) underlie the tree; or indeed with assuming that Root, Trunk and Branch2 (without Branch1) underlie the tree. At the same time it is consistent with assuming that Root and Trunk (without both Branch1 and Branch2) are elements of the tree which *do not* underlie it.

If Root, Trunk, Branch1 and Branch2 are elements of a Unity which is the tree, then none of the molecules which are parts of either of those elements can be elements of the tree. For the molecules are proper parts of proper parts of the tree, and so are pre-elements, but not elements, of the tree. As a consequence, no molecules are such as to underlie the tree. In particular, it is not even the case that *all* the molecules which are parts of the tree at some time *t* (ignoring for the moment the possibility that some molecules are *vaguely* parts of the tree) underlie the tree at that time.

It is consistent with the assumptions that I will be making, however, that those molecules do underlie some other entity which is distinct from the tree - perhaps a lump of wood. For this reason, the theory of Unities as I develop it allows for the superposition of entities, notwithstanding the assumption which I adopt that if any entities underlie a Unity, then that Unity is *unique* (i.e. no other Unities are underlain by the same entities).

The theory of Unities is compatible with assuming that some wholes are not Unities - indeed, that by contrast with Unities, some wholes *are* fundamentally sums. In the version of the theory of Unities I present below, it is assumed that besides Unities there exist *collections* of Unities, where a collection of Unities is itself a Unity if and only if the collection has one member (i.e. itself). I will show that the mereological assumptions regarding such collections combine harmoniously with the very different mereological assumptions made with regard to Unities. It is particularly interesting that the assumptions made regarding collections are formally equivalent to classical mereology. The theory of Unities may be seen, therefore, as a multifaceted theory which includes classical mereology as a component applicable in certain contexts, though not in others.

Looking at the tree again, the theory will acknowledge for example the collection whose members are Root, Trunk, Branch1, and Branch2, as well as a collection whose members are Root and Trunk, or a collection whose members are Branch1 and Branch2. Similarly, there are collections of molecules. Any selected molecules in fact will be the members of a unique collection of molecules, a collection which is their mereological sum. Given that Root, Trunk, Branch1 and Branch2 underlie the tree, I say in this case that the collection of those elements *embodies* the tree. It is clear from what has been said, that I shall be assuming that distinct collections of elements may embody one and the same Unity.

With this introductory sketch of my suggestion in mind, I now turn to a more detailed and critical discussion of its principal components. The desire to show that the conception of wholes as Unities is compatible with a widely accepted core of mereological assumptions has made it necessary to explicate the formal structure of my assumptions. I beg the reader's indulgence with regard to the minuteness of the steps which it has sometimes been necessary to trace. To arrive at an initial, more general impression of the theory of Unities the reader may skip the rest of this chapter, and turn now to Chapter 10.

II - Unities and Collections

1. Collective and Distributive Whole-Parts Relations

Suppose a house has been built solely out of bricks, and that at present *n* bricks, b_1 ... b_n, are parts of the house. Let us imagine that n=10,000. One can say of course that b_1 ... b_n are parts of the house. Furthermore, there is a common use of 'make up' (which I have described as part of the pre-theoretical framework of our discussion in Chapter 4, section I), according to which one can also say that b_1 ... b_n make up the house. According to this use of 'make up', however, it would be denied that b_1 ... b_{10} for example make up the house although they are parts of the house. Similarly, it would be denied that b_3 and b_{42} make up the house although they are parts of the house. Evidently 'the *x*s are parts of *y*' does not mean the same as 'the *x*s make up *y*'.

According to this usage, use of the expressions 'are parts of' and 'make up' is closely bound up with the use of 'is a part of'. The following assumptions can be stated uncontroversially, I think, as constraining the semantic relations between these expressions.

Whereas with respect to 'are parts of' there is the biconditional assumption:

A1 for any *x*s, for all *y*,
the *x*s are parts of *y* if and only if each of the *x*s is a part of *y*

There is, with respect to 'make up', merely the corresponding conditional assumption:

A2 for any *x*s, for all *y*,
if the *x*s make up *y* then each of the *x*s is a part of *y*

Furthermore, while it follows from A1 that[1]

C1 for all *x*s, for all *y*, for all *z*s,
if the *x*s are parts of *y*, and the *z*s are among the *x*s,
then the *z*s are parts of *y*

[1] For proof see Appendix to this chapter (p.249 below).

the example above involving bricks b_3 and b_{42} indicates that regarding 'make up' by contrast one assumes:

A3 for some *xs*, for some *y*, for some *zs*,
the *xs* make up *y*, and the *zs* are among the *xs*,
but the *zs* do not make up *y*

Similarly, it follows from A1 and the transitivity of *is a part of* that[2]

C2 for any *xs*, for all *y*, for all *u*,
if the *xs* are parts of *y*, and *y* is a part of *u*,
then the *xs* are parts of *u*

By contrast, however, considering that bricks might make up a section of a house, without making up the house (of which that section is a part), it is assumed regarding 'make up' that:

A4 for some *xs*, for some *y*, for some *u*,
the *xs* make up *y*, and *y* is a part of *u*,
but the *xs* do not make up *u*

Comparison of assumptions A1 with A2, C1 with A3, and C2 with A4, offers us a clear contrast between ways 'are parts of' and 'make up' are used (according to the familiar usage I am assuming). Taking these assumptions simply to explicate an existing linguistic usage, they may be conveniently referred to as 'the linguistic assumptions'.

In association with the contrast between the ways in which the expressions 'are parts of' and 'make up' are used, a distinction must be drawn between two notions which might be described as notions of a 'whole-parts relation'. If one claims that *y* bears to the *xs* the whole-parts relation, one might either mean simply that the *xs* are parts of *y*, or, more strongly, that the *xs* make up *y*. Since in some contexts it is the case that the *xs* are parts of *y*, though not that they make up *y*, it seems clear that two relations must be distinguished, relations which are ambiguously described by means of the expression 'the whole-parts relation'. In one sense of the latter expression *y* bears the whole-parts relation to the *xs* if and only if the *xs* are parts of *y*. In the other sense *y* bears the whole-parts

---

[2] For proof see Appendix (p.249).

relation to the *xs* if and only if the *xs* make up *y*. It is useful to have distinct expressions to designate these two relations. Let us define therefore the expressions 'is a whole with respect to' and 'is a whole which corresponds to' as follows:

D1 *y* is a whole with respect to the *xs* $=_{\text{def.}}$
the *xs* are parts of *y*

D2 *y* is a whole which corresponds to the *xs* $=_{\text{def.}}$
the *xs* make up *y*

With reference to the fact that the predicate in sentences of the form 'the *xs* are parts of *y*' applies distributively, while the predicate in sentences of the form 'the *xs* make up *y*' applies collectively,[3] the whole-parts relations defined in D1 and D2 by means of these predicates (namely *is a whole with respect to* and *is a whole which corresponds to*) can be referred to respectively as 'the distributive whole-parts' relation and 'the collective whole-parts relation'. By contrast, the relation *is a part of* can be referred to as 'the part-whole relation'. Note that whereas the part-whole relation is dyadic, both the distributive and the collective whole-parts relations are neither dyadic nor even *n*-adic (for some *n*) but rather multigrade relations.

It is noteworthy that the linguistic assumptions (A1-4, C1-2) guarantee (given definitions D1 and D2) that the distributive whole-parts relation can be definitionally reduced to the part-whole relation, but do not similarly guarantee that such a definitional reduction is possible regarding the collective whole-parts relation. To see this, note that it follows from A1 and D1 that *y* is a whole with respect to the *xs* if and only if each of the *xs* is a part of *y*. That is, one might define 'is a whole with respect to' as follows:

D3 *y* is a whole with respect to the *xs* $=_{\text{def.}}$
each of the *xs* is a part of *y*

---

[3] I say that a predicate of the form 'are *F*' in sentences of the form 'the *xs* are *F*' applies *distributively* if sentences of the latter type are equivalently analyzed as the conjunction 'x_1 is *F* and x_2 is *F* and ... and x_n is *F*', where x_1 ... x_n are all and only the *xs*. A predicate of the form 'are *F*' applies *collectively* if sentences of the form 'the *xs* are *F*' cannot be analyzed in the way indicated. The distinction was explained above in the Appendix to Chapter 2.

This means that in formulating a mereological theory which accepted the linguistic assumptions and definitions D1-2, any assumptions adopted regarding the relation *is a part of* will fix at the same time the assumptions adopted regarding the relation *is a whole with respect to*.

By contrast, assumptions which one adopts regarding the relation *is a whole which corresponds to* are relatively independent of those adopted regarding the relation *is a part of*. For 'is a whole which corresponds to' cannot be defined in terms of 'is a part of', unless one makes further assumptions which would connect the relations designated by these expressions - beyond the assumptions discussed so far.[4]

It is, indeed, very tempting to suppose that *is a whole which corresponds to* (no less than *is a whole with respect to*) should be definitionally reducible to *is a part of*, and so to seek further assumptions which would provide for a reductive definition. For it seems natural to suppose that *is a part of* and *is a whole which corresponds to* (as well as *is a whole with respect to*) are somehow correlative to one another, much like the relations *is smaller than or equal to* and *is larger than or equal to*. The sense that the former pair of relations is in some way analogous to the latter pair is perhaps strengthened by the observation that *is a proper part of* is plausibly assumed to be a partial ordering (i.e. a relation which is reflexive, antisymmetrical and transitive), just as much as *is smaller than or equal to* is.

Whether for this reason, or perhaps due to historical connections between mereology and class theory,[5] it seems to be very widely

---

[4] The relation *are parts of, and weigh at least half as much as* complies (as can be seen by inspection) with the assumptions made regarding the relation *make up*, i.e. A2, A3 and A4. However, it is clear that the familiar use of 'make up' which I am discussing here is distinct from that of 'are parts of, and weigh at least half as much as'. Thus assumptions A2, A3 and A4 do not suffice to define 'make up' in terms of 'is a part of'. In particular, given the definition of 'is a whole which corresponds to' in terms of 'make up' (D2), they do not suffice to define 'is a whole which corresponds to' in terms of 'is a part of'.

[5] If one is guided, in developing a theory of wholes and parts, by the theory of classes, one is likely to assume that *make up* is reducible to *is a part of*, and thus that *is a whole which corresponds to* is reducible to *is a part of*. For one might draw a distinction regarding classes which is parallel to the distinction between 'are parts of' and 'make up' discussed above regarding wholes. That is, one might distinguish between 'are members of', on the one hand, and 'form', on the other hand, so that if $s_1 \ldots s_n$ are all the stamps in my stamp-collection, and if C is the class of those stamps, the following claims can be affirmed:

1. $s_1 \ldots s_n$ are members of C
2. $s_1 \ldots s_n$ form C

and

presupposed in current discussions concerning wholes and parts that *is a whole which corresponds to* is adequately accounted for as the relation *is a sum of*:

D4 y is a whole which corresponds to the xs $=_{\text{def.}}$
y is a sum of the xs

Given the conventional definition of 'is a sum of' in terms of 'is a part of', this is to presuppose the following definition of 'is a whole which corresponds to' in terms of 'is a part of' (recall that 'overlaps' is defined in terms of 'is a part of': for u overlaps v if and only if for some w, w is a part of u and w is a part of v)

D4' y is a whole which corresponds to the xs $=_{\text{def.}}$
for all z, z overlaps y if and only if z overlaps one of the xs

Assumption D4 (or D4') is an implicit characteristic of what I have described as 'the conception of wholes as sums'. It appears to be so deeply ingrained that rarely if ever does the possibility surface in contemporary debates, that this assumption might be rejected. I have noted, however, that no reduction of *is a whole which corresponds to* to the relation *is a part of* is entailed by the linguistic assumptions outlined above. Furthermore, the analogy between the pair *is a part of* and *is a whole which corresponds to* and the pair *is smaller than or equal to* and *is larger than or equal to* (an analogy which is likely to suggest that a reduction of *is a whole which corresponds to* should be available in principle) is far from perfect. In particular, whereas both *is smaller than*

---

3. s_2, s_{17}, and s_{225} are members of C
but
4. s_2, s_{17}, and s_{225} do *not* form C
The relation *form*, however, as holding between members (taken together) and classes, is straightforwardly reducible to the relation *is a member of*, given the assumption of extensionality of classes:
the xs form $y =_{\text{def.}}$
for all z, z is one of the xs if and only if z is a member of y
If the theory of wholes and parts is conceived of as closely analogous to the theory of classes (recall that Leśniewski took his mereology to be a theory of what he called 'collective classes'), then one should expect that *make up* (which corresponds to *form* in the theory of classes) be reducible to *is a part of* (which corresponds to *is a member of* is the theory of classes).

or equal to and *is larger than or equal to* are binary relations, *is a part of* - as noted above - is binary but *is a whole which corresponds to* is not (indeed, nor is it *n*-ary, for some *n*; rather, it is a multigrade relation).[6] And the former pair presents no parallel to the distinction between a distributive and a collective whole-parts relation.

It has been shown in Part Two that the conception of wholes as sums involves great difficulties. It would therefore seem to be highly desirable to explore an alternative conception, if such were available. Our discussion indicates a direction in which such an alternative might be sought. A mereological theory might be considered, which rejects definition D4 (as well as D4'). In particular, this involves the proposal that a whole which corresponds to some entities need not be a sum of those entities. Wholes of this sort I shall describe as 'Unities'.

Use of 'Unity' to designate a kind of comprising entity, or a kind of whole, is not unprecedented. For example, Russell uses the term to describe one of the three notions of a whole which he distinguishes in *The Principles of Mathematics*.[7] My use of the term, however, is in no way meant to signify any connection with such former uses.

The term 'Unity' is chosen principally because of what seems to be its intuitive suitability as well as the fact that other candidates either strongly connote notions which need to be dissociated from the notion of a Unity (e.g. 'unit', 'monad', 'singleton', 'integral whole'), or are simply too artificial sounding and awkward (e.g. 'unit-entity', 'uniton').

The theory of Unities will require other specialised terms. Proper parts of Unities are called 'pre-elements'. *Immediate* proper parts of Unities are called elements.[8] A relation designated by 'underlie' plays a role in the theory of Unities which is analogous to the role played by 'compose' in

---

[6] If we describe the converse of the part-whole relation *is a part of* as the whole-part relation (note 'part' in the singular), then this pair of relations, the part-whole relation and the whole-part relation - both being binary relations - is more closely analogous to the pair *is smaller than or equal to* and *is larger than or equal to*.
This analogy, however, only gives grounds for supposing that the whole-part relation is reducible to the part-whole relation, and not for supposing that the collective whole-parts relation is similarly reducible.

[7] See Russell 1903, 139f.

[8] 'immediate proper part' is defined as follows:
x is an immediate proper part of $y =_{\text{def.}}$
x is a proper part of *y* and there is no *z* such that *x* is a proper part of *z* and *z* is a proper part of *y*.
See Simons 1987, 108.

classical and neoclassical mereology. If each of the *xs* is an element of *y*, and *y* is a whole which corresponds to the *xs*, then the *xs* underlie *y*. All this will become much clearer in the course of the systematic presentation of the theory below in section III.

A striking difference between the conception of wholes as Unities and the conception of wholes as sums appears when the following assumption is introduced, which it is possible to adopt consistently only because D4 has been rejected.

A5 for some Unity *y*, for some *xs*, for some *zs*,
y is a whole which corresponds to the *xs*, and
y is a whole which corresponds to the *zs*, and yet
the *zs* are properly among the *xs* (i.e. the *zs* are some
but not all of the *xs*)

Thus if the house for example is a Unity, then it is consistent with A5 that the house is a whole which corresponds to the 10,000 bricks, and yet at the same time a whole which corresponds to 9,999 of the bricks. To assume this would be clearly inconsistent with the conception of wholes as sums, given its adherence to D4 (and D4'). Nothing can be at the same time a sum of the 10,000 bricks and a sum of the 9,999 bricks. For the 10,000th brick overlaps any sum of the 10,000 bricks, and no sum of the 9,999 bricks.[9]

To take another example, suppose Head, Neck, Torso, Leg1, Leg2, Leg3, Leg4 and Tail are parts of a certain cat at some moment of time *t*. Suppose Tibbles is a whole which corresponds to all those parts, whereas Tib is a whole which corresponds to all those parts *except Tail*. If one accepts D4 then one is bound to take Tibbles and Tib to be distinct from one another. If, however, denying D4, one affirms A5, it is open to one to claim that Tibbles is identical to Tib. Section III of Chapter 10 is devoted to discussing a familiar paradox often presented in terms of this example. I argue there that an approach based on the theory of Unities offers a better account of the paradox than do approaches based on the conception of wholes as sums.

---

[9] To verify this claim, suppose that *c* is a sum of $b_1 \dots b_{n-1}$ and *d* is a sum of $b_1 \dots b_{n-1}, b_n$. From the definition of 'sum' it follows that: for all *x*, *x* overlaps *c* iff *x* overlaps one of $b_1 \dots b_{n-1}$; and for all *x*, *x* overlaps *d* iff *x* overlaps one of $b_1 \dots b_{n-1}, b_n$. This entails (assuming b_n overlaps none of the other *bs*) that b_n does not overlap *c*, but that it does overlap *d*.

Even if one has denied D4 (and D4'), and accepted A5, it is tempting to seek alternative ways (replacing D4') in which *is a whole which corresponds to* might be reduced to *is a part of*. One who grants the plausibility of affirming that the house is a whole which corresponds to some 9,999 of its 10,000 bricks would presumably deny that the house is a whole which corresponds, say, to some 20 of its bricks. One would assume that it should be possible to demarcate, at least roughly, the boundary between cases where the *x*s (each being a part of *y*) are such that *y* is a whole which corresponds to them, and cases where they are such that *y* is not a whole which corresponds to them.

One might propose for example that some *x*s are such that the house is a whole which corresponds to them, if and only if the *x*s occupy (collectively) roughly 95% of the region of space occupied by the house. Perhaps other conditions might be added, for example regarding relative proximity of the *x*s.

However, even if one managed to justify the claim that the house is a whole which corresponds to the *x*s if and only if such conditions are fulfilled, it seems unlikely that the same conditions could be justified regarding entities of other types, that is, regarding horses, trees, or stones, instead of houses. Perhaps for stones, for example, one should replace 95% with 60% (in the condition noted in the previous paragraph). Perhaps for horses, and for trees, one should develop conditions which are sensitive to differences between the parts as regards their relative importance for the existence of the whole.

Such reflections make it plausible to assume that having set aside the conception of wholes as sums, no general conditions can be formulated to replace D4' independently of the *sort* of entities concerned. I shall take it as a working assumption, therefore, that no sortally-independent condition is available which would replace D4', and hope to show that this is not a drawback but rather an advantage of the theory of Unities.

With regard to its rejection of D4' and of any alternative (sortally independent) definition of 'is a whole which corresponds to' in terms of 'is a part of', the theory of Unities is appropriately described as being *holistic*, in a sense in which no theory which embraces the conception of wholes as sums can be. For whereas a theory of the latter type takes the notion of a whole to be conceptually derivable from the notion of a part (for both 'is a whole with respect to' and 'is a whole which corresponds to' are definable in terms of 'is a part of'), the theory of Unities takes the notion of a whole not to be so derivable (for although 'is a whole with respect to' is definable in terms of 'is a part of', 'is a whole which corresponds to' is not).

2. Wholes and Sums

To reject the conception of wholes as sums is not the same as rejecting the claim that all wholes are sums. By rejecting D4, one is rejecting the claim that a whole which corresponds to the *xs* is necessarily a sum of the *xs*. One is not rejecting the claim that a whole which corresponds to the *xs* is necessarily a sum of *some zs*. To be clearer on this point, let us distinguish between saying that an entity is *fundamentally* a sum, and saying that it is *peripherally* a sum. An entity is peripherally a sum if, simply, there are some *xs* of which it is a sum:

D5 *y* is peripherally a sum $=_{def.}$
for some *xs*, *y* is a sum of the *xs*

By contrast, an entity is fundamentally a sum if in addition to being peripherally a sum, it is such as to be a sum of any *xs* to which it is a corresponding whole:

D6 *y* is fundamentally a sum $=_{def.}$
1. for some *xs*, *y* is a sum of the *xs*
2. for any *xs*, if *y* is a whole which corresponds to the *xs*, then *y* is a sum of the *xs*

It is entailed by the conception of wholes as sums, given its acceptance of D4, that all wholes are fundamentally sums. Indeed, it is because this conception is committed to the claim wholes are fundamentally sums, rather than the weaker claim that wholes are peripherally sums, that it is particularly appropriate to describe this conception as the conception of wholes as sums.

The theory of Unities, in as much as it accepts assumption A5, must reject the claim that all wholes are fundamentally sums since this claim is clearly inconsistent with A5. It is consistent with A5, however, to accept the claim that all wholes are peripherally sums. Moreover, if one assumes that *is a part of* is transitive, one is bound to accept that if the *xs* are strictly *all* the parts of *y* then *y* is a sum of the *xs*. Both because I shall be accepting the transitivity of *is a part of* in what follows, and because I take it to be intrinsically reasonable to claim that a whole must be a sum of (strictly) all its parts, I shall assume that all wholes are peripherally sums, whether or not they are Unities:

A6 for all *x*, if *x* is a whole than *x* is peripherally a sum

3. Classical, Neoclassical, and Unity-theoretical Assumptions

It is useful to summarise the foregoing introduction to the notion of a Unity by comparing the constraints on relations between wholes and sums, as assumed in classical mereology, in neoclassical mereology, and in the theory of Unities.

There are three noteworthy features of wholes as conceived of in classical mereology: Universal existence, uniqueness, and coextensivity of parts. These features have been discussed in Part Two, and may be explained now using the terminology introduced above as follows:

<u>universal existence</u>:
wholes are universally existent if and only if for any *x*s, there is a whole which corresponds to those *x*s

<u>uniqueness</u>:
wholes are unique if and only if for any *x*s, for all *y*, for all *z*, if *y* is a whole which corresponds to the *x*s and *z* is a whole which corresponds to the *x*s, then *y* is identical to *z*

<u>coextensive determination</u>:
wholes are coextensively determined if and only if for all *u*, for all *v*, for any *x*s, for any *y*s, if *u* is a whole which corresponds to the *x*s, and *v* is a whole which corresponds to the *y*s, then *u* is identical to *v* only if the *x*s are coextensive with the *y*s[10]

Neoclassical mereology rejects either universal existence, or uniqueness, or both. However, both classical mereology and neoclassical mereology are compelled to maintain coextensivity of parts, since this follows from the way the term 'sum' is defined (as explained in section I of Chapter 8), and from acceptance of definition D4, whereby wholes are made out to be fundamentally sums.

---

[10] As noted in Chapter 3, section III, the *x*s are coextensive with the *y*s if and only if for all *z*, *z* overlaps one of the *x*s if and only if *z* overlaps one of the *y*s.

The theory of Unities differs from both classical and neoclassical mereology in rejecting coextensivity of parts. It is able to do so by rejecting the claim embodied in D4, according to which wholes are fundamentally sums. As usual with neoclassical mereology, and in contrast with classical mereology, the theory of Unities rejects the universal existence of wholes. As for uniqueness, its position might be roughly described as midway between classical and neoclassical mereology. A precise explanation of this issue, however, is best left to the systematic presentation in the next section.

4. Unities and Collections

Although it is of great interest to investigate possible applications of the theory of Unities *outside* the context of the study of concrete entities,[11] I shall for the purposes of the present work confine myself to concrete entities. It will be useful at this stage to state briefly the chief intended interpretation of the central term of the theory, namely 'Unity', to facilitate a less abstract discussion: I take the term 'Unity' to apply to concrete *individuals*. Furthermore, I propose the conception of such individuals as Unities as a comprehensive alternative to their prevalent conception as a neoclassical mereological sum.

It is noteworthy however that given this interpretation, the theory of Unities is compatible with the view that not all concrete entities are individuals. In particular, it will be shown in detail in the next section that the theory of Unities is compatible with the view that for any Unities, there exists a *collection* of those Unities. Such a collection is itself a Unity if, and only if, it has *only one member*. Since a multi-membered collection of Unities is not itself a Unity, and individuals are assumed to be Unities, it is compatible with the theory that entities - multi-membered collections - exist, which are not individuals.

The notion of a collection in this context is understood in accordance with the account offered in Part One. It was suggested that the notion of a collection is indeterminate in various respects (for example, in respect of the requirement that a collection should be unified, or that it be monadic, or that the relation *is a part of*, as applying to collections, be transitive). It transpired in Part Two that the notion of a classical mereological sum is one plausible determination of the notion of a collection. It is this

[11] A comment to this effect was mentioned in the Introduction to this work.

determined notion of a collection that I adopt in the present context, in supplementing the theory of Unities with a theory of collections.

Therefore, the theory of collections presented here will have formal features which precisely parallel the formal features of classical mereology. Thus if a_1, a_2, .. , a_6 are Unities, and A is a collection of these Unities, A is naturally described as a classical mereological sum of a_1, a_2, .. , a_6, with only one qualification: A is not itself an individual, for it is a multi-membered collection. Traditionally, classical mereology is assumed to be a calculus of *individuals*, where the classical sum of any individuals is itself an individual. Although the theory of collections presented here is formally indistinguishable from classical mereology, it cannot be described as a calculus of *individuals*, but rather as a calculus of *collections* of individuals (i.e. collections of Unities).

This difference, however, does not arise from any formal difference between the theories, but rather from a difference in the way the terms of the theories are respectively interpreted. On the traditional interpretation of classical mereology as a calculus of individuals, the variables of the theory range over individuals; on the present interpretation of classical mereology as a calculus of collections, the variables of the theory range over collections of Unities.

To supplement the theory of Unities with such a theory of collections is not to concede that the theory of Unities presupposes classical mereology. The assumptions required for a theory of Unities are presented in the next section separately from the supplementary assumptions required for a complementary theory of collections, and thus it will be clear that the latter theory is not presupposed by a theory of Unities. In particular, classical mereology is not presupposed.

However, given that classical mereology is an internally consistent theory,[12] and given that on the currently intended interpretation one is not committed to the existence of arbitrary sum-*individuals*, nor even, as we shall see, to the view that distinct individuals cannot be at the same place at the same time, there seems to be little harm in accepting classical mereology thus interpreted.

On the other hand, accepting the existence of collections, and in particular collections understood as classical sums, is in my view both inherently plausible and convenient. The existence of such collections would provide a straightforward explanation of, and justification for, our

---

[12] See Simons 1987, 110, n. 23.

talk of such entities as a collection of pebbles or a group of dots on the blackboard. Furthermore, accepting the existence of such collections in the present context allows us to demonstrate how a well established traditional theory such as classical mereology can be smoothly combined with assumptions about Unities.

To obtain a clearer first impression of the ontological picture that results, consider again our example of the house *h* and its *n* bricks, the *b*s. In order to have concrete numbers in mind, suppose *n*=100. Suppose each of the bricks is a Unity, and the house itself is a Unity. The assumptions so far do not entail that there are more than 101 Unities. Each of these Unities is a single-membered collection. However, the assumptions do entail that there is an unimaginably larger number of (non-single-membered) collections. For any two bricks, there is a collection of those bricks; and similarly for any three bricks, four bricks, ... , *n* bricks. To be precise, the number of collections of bricks works out to be 2^{100} - 1. That is to say, there are roughly 1,267 billion billion billion distinct collections of bricks.

Given the result of our discussion in Chapter 8, however, any of these numerous collections is collectively identical to the bricks that compose it. This means that the assumption that such collections exist can be said in the strictest sense to involve no extension of our ontological commitments. Granted the existence of the 101 Unities, the assumption that arbitrary sums of those Unities exist introduces no new entities, that is, no entities that are not collectively identical to some or all of the 101 existing Unities.

A collection, being an entity which complies with classical mereological assumptions, can be described (in terms explained in the former section) as being *fundamentally* a sum. A Unity, by contrast, is only *peripherally* a sum. Now suppose the *a*s are some, but not all of the bricks (i.e. suppose the *a*s are properly among the *b*s), and suppose *A* is the collection of the *a*s and *B* is the collection of the *b*s. It follows that *A* is a whole which corresponds to the *a*s but not to the *b*s, whereas *B* is a whole which corresponds to the *b*s but not to the *a*s. The house, however, since it is assumed to be a Unity, may be (though of course need not be) at the same time a whole which corresponds to the *a*s and to the *b*s.

Let us call the collection of the *b*s the *complete* collection of bricks (at time *t*). It is one of the noteworthy features of the theory of Unities that it allows us to draw a clear distinction, in a case such as this, between the house and the complete collection of bricks that underlie it at *t*, irrespective of any assumptions regarding the disparity of conditions under which the house and the collection might survive, respectively.

The distinction between Unities and multi-membered collections is a distinction between two fundamentally different notions of a whole. These notions need not be viewed as competing with one another, but rather as complementing one another. As will emerge more clearly from the systematic presentation of assumptions in the next section, a collection is a whole which is identical to its parts. A Unity, on the other hand, is a whole which is not identical to its parts.

Both Unities and multi-membered collections are *single* entities. It was argued in Part One that singleness does not conflict with identity to the many parts. However, in that a Unity is not identical to its parts, nor, indeed, to any *x*s, if the *x*s are greater than 1 in number, Unities are *monadic* entities.[13] Multi-membered collections, by contrast, are *non-monadic* entities. Entities of both types have parts. But the character of the respective relations between these entities and their parts is fundamentally different. On my analysis, the main difference lies in the difference between the species of the relation *is a whole which corresponds to* which apply respectively to collections and Unities, and in the attendant difference between entities that are fundamentally sums and entities that are only peripherally sums.

It is true that according to the account I have offered, neoclassical sums, too, are wholes conceived of as non-identical to their parts (see relevant comments in Chapter 8, section II). However, we have seen in Chapter 6 that the notion of a neoclassical sum is deeply problematic, and if it has to be rejected, it is important to realise that we have an alternative conception of a whole, one which also requires the conception of a whole as non-identical to its parts. Moreover, the notion of a Unity goes much further than the notion of a neoclassical sum in distinguishing itself from the contrasting notion of a whole which is identical to its parts. The grounds offered by neoclassical mereology for distinguishing between a whole and its parts depend on extra-mereological considerations, such as the survival of individuals under circumstances of flux of parts and of changes in the relations between parts. The theory of Unities provides us in addition with intrinsic mereological grounds for distinguishing between a whole and its parts (if the whole is a Unity).

---

[13] See Chapter 2 section II.

Furthermore, neoclassical mereology seems unable to offer a conception of a whole which is (in terms of the discussion of section II of Chapter 7) presently discernible from a superposed classical sum. By contrast, it will be seen that a Unity *is* presently discernible from a superposed classical sum. Thus, although neoclassical mereology proposes wholes which are not identical to their parts, these are wholes which are at all times (presently) indiscernible from classical sums, and thus they are at all times (presently) indiscernible from all the parts (taken collectively). This is not a compelling notion of a whole which is not identical to its parts. The contrast offered by the theory of Unities between wholes which are identical to their parts and wholes that are not, I suggest, is much starker.

III - Principles for a Theory of Unities

I now turn to show how basic principles may be constructed for a theory of wholes and parts in which the notions of a Unity and of the relation *underlie* play a central role. What I offer here cannot, by any means, be viewed as a fully developed theory. It will be evident that a theory based on the core notion of a Unity may be developed in many and varied ways. My aim is not to survey all the possibilities that arise, but rather to present an initial picture which can serve as an indication of the way in which a more complete and well rounded theory could be developed. In systematic terms, the principles I have collected are rather loosely organised. I have attempted to put together a picture which makes what I take to be important features of the theory clear, rather than one which emphasizes the inferential relations governing the system as a whole. In particular, I make no claim that the version I present cannot be shown to be founded on a smaller number of principles, from which some of the others could be derived.

As before, I leave the temporal modification of predicates such as 'is a part of', 'is coextensive with', 'is a sum of', 'underlie', etc. almost completely implicit, to simplify the presentation. Supplementing the account with explicit temporal modification of these predicates presents no problems of principle.

1. Preliminary Assumptions Regarding Unities and Collections

We begin by assuming that some entities are Unities, viewing them as belonging to a corresponding ontological category, the category of Unities. That is,

A1 for some *x*, *x* is a Unity

If *x* is a Unity, then *x* is a single entity, in the sense explained in Part One. In that discussion it was urged that the claim that *x* is a single entity is compatible with the claim that *x* is (collectively) identical to many entities. However, we shall assume that a Unity cannot be identical to many Unities (nor to many entities of other types). To describe this feature of Unities, we may use the term 'monadic', defining it as follows:

D1 *y* is monadic $=_{def.}$
for any *xs*, *y* is identical to the *xs* only if the *xs* are 1 in number

With regard to Unities, therefore, we assume:

A2 for all *x*, if *x* is a Unity then *x* is monadic

It would seem that the only context in which it might be plausible to suggest that a single entity is identical to many distinct entities is a case in which the former is a whole which corresponds to the latter.[14] In order to allow, as we shall be doing, that a Unity *u* might both be a whole which corresponds to the *xs* and a whole which corresponds to the *ys*, even though the *xs* are not coextensive with the *ys*, we would seem to be compelled at least with regard to those cases to assume that *u* is neither identical to the *xs* nor to the *ys*.

There are other reasons for assuming A2, some deriving from the mereological assumptions presented below, others from considerations

---

[14] The claim that a whole is identical to all its parts is not without its sympathisers, as we have seen in Chapter 8. I am not aware, however, that the claim that a single entity is identical to many entities (taken collectively) has ever been put forward with regard to concrete entities in connection with cases in which the many entities are not taken to be *parts* of the relevant single entity (though perhaps such a claim is made with regard to universals by some of the theorists who conceive of universals as repeatables - see Armstrong 1989; Bigelow 1988).

associated with the intended interpretation of Unities as individuals.[15] Examining the details of these considerations would lead us far astray, however, and even if restricting the scope of A2[16] is compatible with other requirements of a theory of Unities, the discussion is simplified by adopting A2 in its present unrestricted form.

Next, we introduce collections of Unities. For some entities, the *xs*, we assume that there exists a unique entity which is a collection of the *xs*. Thus we assume:

A3 for some *u*, and for some *xs*, *u* is a collection of the *xs*

A4 for all *u*, for all *v*, and for any *xs*, if *u* is a collection of the *xs* and *v* is a collection of the *xs*, then $u = v$

Collections we take to be always collections of Unities. And such collections exist universally. That is,

A5 for all *u*, and for any *xs*, if *u* is a collection of the *xs* then each of the *xs* is a Unity

A6 for any *xs*, if each of the *xs* is a Unity, then for some *y*, *y* is a collection of the *xs*

Suppose three bricks b_1, b_2, and b_3 taken collectively are a value of the variable '*xs*' (note that because '*xs*' is a plural variable, it may take as *one* of its values more than one entity). Then on our assumptions, if there is a collection of the *xs* when the variable takes b_1, b_2, and b_3 (collectively) as its value, then each of the bricks is a Unity. Conversely, if each of the bricks is a Unity, then there exists a collection of the *xs*, when '*xs*' takes b_1, b_2, and b_3 (collectively) as its value.

Similarly (noting that although a plural variable may take more than one entity as its value, it may also take only one entity as its value), if there is a collection of the *xs* when '*xs*' takes only one of the bricks, say b_1, as its value, then that brick is a Unity. And conversely, if b_1 is a Unity,

---

[15] Similar considerations lead to the conclusion that a neoclassical conditioned sum, for instance, is not identical to its parts. An example of such a consideration is the observation that in some cases the parts of an individual survive the individual's destruction.

[16] E.g. assuming A2 only in cases where *x* has more than two elements.

then there exists a unique collection of the *xs*, when '*xs*' takes b_1 as its value.

The notion of a member is defined as follows:

D2 *y* is a member of $u =_{def.}$
for any *xs*, if *u* is a collection of the *xs* then *y* is one of the *xs*

It follows that[17]

C1 for any *ys*, for all *u*, for any *xs*,
all and only the *ys* are members of *u* if and only if
if *u* is a collection of the *xs* then the *ys* = the *xs*

The notions of a sub-collection and a proper sub-collection are defined as follows:

D3 *u* is a sub-collection of $v =_{def.}$
for all *x*, *x* is a member of *u* only if *x* is a member of *v*

D4 *u* is a proper sub-collection of $v =_{def.}$
u is a sub-collection of *v* and for some *x*, *x* is a member of *v* but not of *u*

We assume that a single-membered collection is identical to its member:

A7 for all *u* and for all *x*, if *u* is a collection whose only member is *x*, then $u = x$

Assumption A7 guarantees that a single-membered collection is a Unity:[18]

C2 for all *u*, if *u* is a single-membered collection then *u* is a Unity

From A6 and A7, we may infer:[19]

C3 for all *u*, *u* is a Unity if and only if it is a

---

[17] For proof see Appendix (p.250).
[18] For proof see Appendix (p.251).
[19] For proof see Appendix (p.251).

single-membered collection

However, this still does not preclude a many-membered collection from being a Unity, for our assumptions so far are consistent with a single-membered collection being identical to a many-membered one. We rule out this possibility, and simplify the relations between Unities and collections, by assuming the following:

A8 for all *u*, if *u* is a collection with more than one member, then *u* is not a Unity

If there are at least two Unities then C3, A6, and A8 guarantee that the ontological category of Unities is strictly a sub-category of the category of collections.

2. Mereological Assumptions Regarding Unities and Collections

We lay down as a foundation for a mereological theory involving the notion of a Unity definitions and assumptions which demonstrate the wide common ground between this theory, on the one hand, and classical and neoclassical mereology on the other.

Treating 'proper part' as a primitive term, we define 'improper part', 'part', 'immediate part', 'overlap', 'disjoint' and 'sum' in agreement with common mereological usage, adding to this list the term 'coextensive', of which we have already made use in the discussion above. Note that although the theory of Unities is compatible with the phenomenon of being in the same place at the same time, we do need not resort to the rather extreme measure which Simons requires for this purpose, that of denying D5, an assumption maintained both in the classical mereological view and, it would seem, by common sense.[20]

D5 *x* is a part of *y* $=_{\text{def.}}$
x is a proper part of *y* or *x* is identical to *y*

D6 *x* is an improper part of *y* $=_{\text{def.}}$
x is a part of *y* and *x* is not a proper part of *y*

[20] See Simons' discussion of issues involved here at Simons 1987, 112f.

D7 *x* is an immediate proper part of *y* $=_{def.}$
x is a proper part of *y* and it is not the case that
for some *z*, *x* is a proper part of *z* and *z* is a proper part of *y*

D8 *x* overlaps *y* $=_{def.}$
for some *z*, *z* is a part of *x* and *z* is a part of *y*

D9 *x* is disjoint from *y* $=_{def.}$
it is not the case that *x* overlaps *y*

D10 *y* is a sum of the *xs* $=_{def.}$
for all z, z overlaps *y* if and only if *z* overlaps at least one of the *xs*

D11 the *xs* are coextensive with the *ys* $=_{def.}$
for all z, z overlaps at least one of the *xs* if and only if *z* overlaps at least one of the *ys*

Next, we lay down widely accepted assumptions governing the relation *is a proper part of*, expressing the asymmetry and transitivity of the relation,[21] as well as its compliance with the Principle of Supplementation of Proper Parts:[22]

---

[21] See Simons 1987, 26-28; 362. I have commented on several opportunities above on the controversiality of the assumption of transitivity of *is a proper part of* (A8). The theory of Unities does not depend on making this assumption. However, since it is compatible with this assumption, and since the assumption is characteristic of both classical and neoclassical mereology, it helps to demonstrate the common grounds between the latter theories and the theory of Unities if the assumption is kept.

[22] The Principle of Supplementation of Proper Parts (which Simons calls the Weak Supplementation principle) places a somewhat unnatural constraint on the theory of Unities, in that it implies that no Unity can exist which has one and only one element at some time *t*. Simons comments that Brentano denied this principle (and that a somewhat similar view to Brentano's is held by Kit Fine). However, he asks,

How could an individual have a *single* proper part? That goes against what we mean by 'part'. An individual which has a proper part needs other parts in addition to *supplement* this one to obtain the whole. (Simons 1987, 26).

Simons' point can be countered by arguing that the intuition that a whole must be in some way 'more' than anything that is its proper part is satisfied by the observation that the whole simply has more parts than any of its proper parts do. If *y* has a proper part *x*, we need not assume that *y* has an additional proper part *z* which is disjoint from *x*, to

A9 for all x, for all y, if x is a proper part of y then y is not a proper part of x

A10 for all x, for all y, for all z, if x is a proper part of y and y is a proper part of z then x is a proper part of z

A11 for all x, for all y, if x is a proper part of y then for some z, z is a proper part of y and z is disjoint from x

In addition to assumptions regarding the relation *is a proper part of*, the discussion of section II above has indicated the importance of making explicit our assumptions about the relation *is a whole which corresponds to*. In classical and neoclassical mereology this relation is taken to be accounted for as the relation *is a sum of*, and since the latter relation is analyzable in terms of the relation *is a proper part of* (see definitions D5, D8 and D10 above), the need does not arise to treat 'whole' as a further primitive term of the system, in addition to 'proper part'. However, since in disagreement with the view implicit in classical and neoclassical theories we reject the account of wholes as sums, we are obliged, in effect, to take 'whole' as primitive, no less than 'proper part'.

We accept, of course, that there are analytic connections between 'whole' and 'proper part'. Assumption A12 below, for example, seems to be the expression of such connections. However, we deny that these analytic connections between the terms suffice to establish a relation of interdefinability between them. And therefore the assumptions made above regarding the relation *is a proper part of* do not determine the assumptions adopted regarding the relation *is a whole which corresponds to*. We therefore introduce the following assumptions, which are motivated by our discussion of section II above.

A12 for all y, for any xs, if y is a whole which corresponds to the xs then each of the xs is a part of y

A13 it is not the case that for all y, for any xs, for any zs,

---

guarantee that y has more proper parts than x does. After all, transitivity alone guarantees that y has more proper parts than x does - y has all the proper parts which x does, and in addition it has x itself as a proper part (while x, of course, does not have itself as a proper part). However, for a reason similar to the one noted in the preceding footnote, I have chosen to maintain the Principle of Supplementation of Proper Parts in the present context.

if *y* is a whole which corresponds to the *xs* then if the *zs* are among the *xs* then *y* is a whole which corresponds to the *xs*

A14 for all *y*, for any *xs*, if *y* is a whole which corresponds to the *xs* then for some *zs*, the *xs* are among the *zs* and *y* is sum of the *zs*

Definitions D5-D11 and assumptions A9-A14 can be taken to specify basic features of the relations *is a proper part of* and *is a whole which corresponds to*. This does not mean, however, that we take there to be only one type of relation which may legitimately be described by 'is a proper part of' (nor only one which may be described by 'is a whole which corresponds to'). We may take it that these relations are related to others as generic to specific. That is to say, there is the generic *is a proper part of* (and *is a whole which corresponds to*) which is found in various species. The species of *is a proper part of* share with the genus the formal features outlined above. There may, however, be formal differences between different species of the generic relation, differences which are compatible with the shared core of features.

We will assume, in particular, that there are two distinct species of *is a proper part of*, and correspondingly of *is a whole which corresponds to*. One species of *is a proper part of* is called 'is a pre-element of', and is a relation which holds between *x* and *y* only if both *x* and *y* are Unities. The corresponding species of *is a whole which corresponds to* is called 'is underlain by'. A second species of *is a proper part of* is called 'is a proper sub-collection of', and is a relation which holds between *x* and *y* only if *y* is not a Unity (recall that we are assuming that all concrete entities fall under the category of collections, and that some of them fall under the sub-category of Unities).

The relationship between *is a proper part of*, on the one hand, and *is a pre-element of* and *is a proper sub-collection of*, on the other hand, may be illustrated by means of the analogous relationship between various relations of similarity. Consider the generic relation *is perfectly similar in some respect to*, which can be expressed by means of 'is PS-X to'. The relations *is perfectly similar in respect of colour to*, *is perfectly similar in respect of shape to*, are among the species of this relation - expressed in abbreviated form by 'is PS-colour to' and 'is PS-shape to', respectively. In general, species of *is PS-X to* are designated by expressions of the form 'is perfectly similar in respect of A to', where 'A' is substituted by abstract nouns from a range to which 'colour' and 'shape', among others, belong.

The task of constructing a formal theory of the relation *is PS-X to* is ambiguous between the task of characterizing the generic formal features of species of this relation, and the task of characterizing the formal features of the generic relation itself. On the former understanding, one would lay down the assumption that *is PS-X to* is transitive (for any species of this relation is transitive). On the latter understanding, by contrast, one would lay down the assumption that *is PS-X to* is non-transitive (for the generic relation itself is not transitive).

There is an analogous ambiguity regarding the task of constructing a formal theory of the relation *is a proper part of*. Assuming that this relation is generic, and that at least two relations belong to it as species (namely, *is a pre-element of* and *is a sub-collection of*), then the task might be understood as that of characterizing generic formal features of the species, or as that of characterizing formal features of the generic relation itself. I shall understand the task in the former way.

Thus it becomes clear how one may consistently claim, for instance, that the relation *is a proper part of* is transitive (in the sense that any species of the relation is transitive), and yet assert that although x is a proper part of y, and y is a proper part of z, x is nevertheless not a proper part of z. For if x is a pre-element of y then x is a proper part of y (the former relation is a species of the latter). And if y is a sub-collection of z then y is a proper part of z (again, the former is a species of the latter). But from these assumptions it neither follows that x is a pre-element of z, nor that x is a sub-collection of z. And since these are the only two species of the relation *is a proper part of* which we have held to exist, it does not follow from our assumptions that x is a proper part of z (even though we assert the transitivity of *is a proper part of* in the sense explained).

When we compare the account of the mereology of individuals offered by the theory of Unities with accounts offered by classical mereology, or neoclassical mereology, we may take these theories as sharing the above stated assumptions and definitions, with the exception of definition D5 (this definition is accepted in classical mereology and in the theory of Unities, but not generally accepted in neoclassical mereology). Keeping this difference in mind, we may use the terms involved in these assumptions and definitions in a way relatively free from controversy, and point out particularities of the theory of Unities by using terms specific to it, terms which are associated with further assumptions and definitions to which we now turn.

3. Collections and the Relation *is a proper sub-collection of*

We begin by treating the first of the two species of *is a proper part of* (and of *is a whole which corresponds to*) which we distinguish in the theory of Unities: The relation *is a proper sub-collection of.*

We note first that the formal relation between *is a proper sub-collection of* and *is a sub-collection of* corresponds precisely to that between *is a proper part of* and *is a part of*, respectively, as stated in definition D5. We may further define the relations 'collection-overlaps', 'is collection-disjoint from', and 'is a collection-sum of', in terms of 'is a proper sub-collection of', precisely in parallel to the definitions of 'overlaps', 'is disjoint from' and 'is a sum of', respectively, in terms of 'is a proper part of'.[23] Using these definitions, we can prove on the basis of A3-A7 and D2-D4 that collections form a system with respect to the relation *is a proper sub-collection of* which is isomorphic to classical mereology. For we can prove the following statements, which correspond to the axioms Simons gives for classical mereology:[24]

A15 for all *x*, for all *y*, if *x* is a proper sub-collection of *y* then *y* is not a proper sub-collection of *x*

A16 for all *x*, for all *y*, for all *z*, if *x* is a proper sub-collection of *y* and *y* is a proper sub-collection of *z* then *x* is a proper sub-collection of *z*

A17 for all *x*, for all *y*, if *x* is a proper sub-collection of *y* then for some *z*, *z* is a proper sub-collection of *y* and *z* is collection-disjoint from *x*

A18 for any *x*s, for some *y*, *y* is a collection-sum of the *x*s

---

[23] Indeed, one may supply the theory of collections with definitions which parallel all the definitions given in association with 'is a proper part of' above. However, this will not be required for our purposes; we may always use the terms defined in terms of 'is a proper part of', since the context will generally make it clear in what follows which species of the generic relation is intended.

[24] See Simons 1987, 37, axioms SA1, SA2, SA3 and SA24. For proofs of A15-A18 see Appendix (pp.251-254).

This formal isomorphism with classical mereology allows us to assume, consistently with our assumptions regarding *is a proper part of*, that *is a proper sub-collection of* is a species of *is a proper part of* (a species which is found only in cases where, if *x* is a proper part of *y*, *y* is not a Unity, for a member of a sub-collection does not have proper sub-collections[25]):

A19 for all *y*, for all *x*, if *y* is not a Unity, then
x is a proper part of *y* if and only if *x* is a proper
sub-collection of *y*

As for the species of *is a whole corresponding to*, associated with this species of *is a proper part of*, we make the following assumption, in the spirit of classical (and indeed neoclassical) mereology:

A20 for any *x*s, for all *y*, if the *x*s are sub-collections of *y*, then *y* is a whole which corresponds to the *x*s if and only if *y* is a collection-sum of the *x*s

In effect, A19 and A20 guarantee that our theory of collections is not merely isomorphic to classical mereology, it is an *instance* of classical mereology. It is classical mereology viewed as applying to collections of Unities (which are taken to be collections of individuals) rather than to individuals. This allows us to apply here the result of our discussion of section II of Chapter 8, regarding the identity between a classical sum and its parts.

A21 for all *y*, for any *x*s, if *y* is a collection-sum of the *x*s, then *y* is collectively identical to the *x*s

It follows from A20 and from the isomorphism with classical mereology that anything which is a whole (of this species) which corresponds to the *x*s, is *unique*:

---

[25] A member of a sub-collection is, by A5, a Unity. A Unity is a single-membered collection (by A7). A single-membered collection cannot have proper sub-collections (by D4).

A22 for any *xs*, for all *y*, for all *z*, if the *xs* are sub-collections of *y* and *y* is a whole which corresponds to the *xs*, and if the *xs* are sub-collections of *z* and *z* is a whole which corresponds to the *xs*, then $y = z$

4. Unities and the Relation *is a pre-element of*

We now lay down assumptions associated with the second (and more important) species of the relation *is a proper part of* which is acknowledged by the theory of Unities. On our assumptions so far, no allowance was made for proper parts of Unities (see A19). In the theory of Unities it is assumed, however, that not only multi-membered collections but also Unities may have proper parts. We shall assume that only Unities can be proper parts of Unities. The species of *is a proper part of* which is found as a relation between Unities, however, has formal characteristics distinct from those of the relation *is a proper sub-collection of* (apart from the common core which they share, as expressed by D5-D11 and A9-A14). We therefore designate it by a distinct term, 'is a pre-element of'.[26] Thus, the proper parts of a Unity are its pre-elements.[27]

D12 *x* is a pre-element of $y =_{\text{def.}}$
x and *y* are Unities, and *x* is a proper part of *y*

A particularly important role in the theory of Unities is accorded to *immediate* proper parts of Unities (see definition D7 above). We call such parts 'elements'. Thus we have

D13 *x* is an element of $y =_{\text{def.}}$
x is a pre-element of *y* and *x* is an immediate proper part of *y*

[26] It is possible to present in connection with 'is a pre-element of' a battery of definitions paralleling those given above in association with 'is a proper part of'. This would require introducing many additional terms, which would distinguish between generic relations (associated with 'is a proper part of') and specific relations (associated with 'is a pre-element of'). However, this would be of little use in practice, and no confusion should arise in the context from use of the respective generic terms.

[27] We are not assuming that *all* Unities have proper parts. If a Unity has no proper parts then, of course, it has no pre-elements, and it may be described as an *atomic* Unity.

If each of the *x*s is an element of Unity *y*, and *y* is a whole which corresponds to the *x*s, then the *x*s are said to underlie *y*:

D14 the *x*s underlie $y =_{\text{def.}}$
each of the *x*s is an element of *y*, and *y* is a whole which corresponds to the *x*s

Just as for any *x*s only one collection can be the collection-sum of the *x*s (if each of the *x*s is a collection) - for the theory of collections presented here is, as explained above, an instance of classical mereology, and uniqueness of sums is a feature of classical mereology - we similarly assume that only one Unity can be underlain by the *x*s (if the *x*s are elements of a Unity):

A23 for any *x*s, for all *y*, for all *z*, if the *x*s underlie *y* and the *x*s underlie *z* then $y = z$

However, although we are assuming that for any *x*s, at most one Unity is underlain by the *x*s, we are not assuming that at most one Unity is a whole which corresponds to those *x*s. To bring this point out, consider the following illustration:

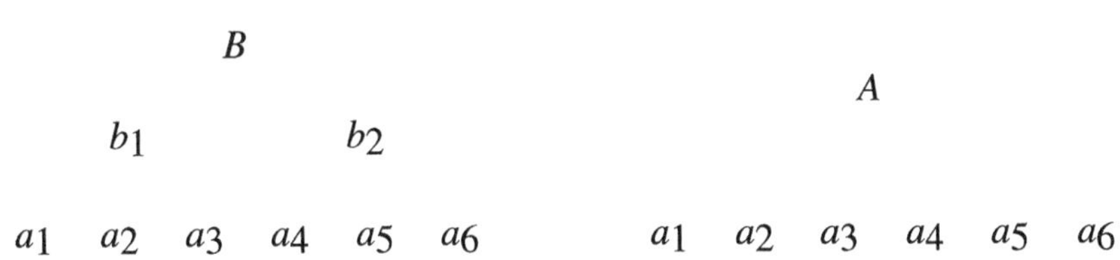

Figure 1

Suppose *B* is an organism (a greatly simplified one); b_1 and b_2 are the organs of *B* (at time *t*) - where entities such as a heart or a muscle are taken to be organs of an organism; and a_1 ... a_6 are the cells of *B* (at time *t*). Suppose we find it reasonable to apply the theory of Unities so as to assume that b_1 and b_2 underlie *B*, and that a_1, a_2 and a_3 underlie b_1, whereas a_4, a_5 and a_6 underlie b_2.

The cells do not underlie B, on these assumptions, for since each of them is a proper part of an element of B (i.e. one of the organs), none of them is an immediate proper part of B, and therefore none of them is an element of B (see D13 and D14). But our assumptions are compatible with the highly plausible supposition that B is nevertheless a whole which corresponds to a_1, a_2, a_3, a_4, a_5, and a_6. If that is the case, then B is both a whole which corresponds to b_1 and b_2, and a whole which corresponds to a_1, a_2, a_3, a_4, a_5, and a_6.

Now, although as just noted it follows from our assumptions that the cells a_1, a_2, a_3, a_4, a_5, and a_6 do not underlie B, it is consistent with our assumptions that there is some Unity A (see illustration) which is not identical to B which the cells a_1, a_2, a_3, a_4, a_5, and a_6 *do* underlie (it might be suggested for instance that A is a lump of flesh which constitutes the organism and yet is not identical to it). If the cells underlie A then by D14 it follows that A is a whole which corresponds to those cells. We conclude, then, on these assumptions, that both A and B are wholes which correspond to a_1, a_2, a_3, a_4, a_5, and a_6, and yet $A \neq B$.

Thus, although for any *x*s no more than one Unity may be underlain by the *x*s, more than one Unity may be a whole which corresponds to the *x*s.

In a fully developed theory of Unities, the relation *is a whole which corresponds to* should be explicated in detail, not only as obtaining between a Unity and its elements (in which case we have a relation between a whole and parts which underlie it), but also as obtaining between a Unity and pre-elements which are *not* elements (in which case we have a relation between a whole and parts which do *not* underlie it). However, it would exceed the scope of the present work to provide such an exposition here.

The relation *is a whole which corresponds to* as obtaining between Unities and their elements is distinguished from the relation as obtaining between collections and their sub-collections, principally in that while the latter requires that wholes be fundamentally sums, the former does not. Even if the *x*s are not coextensive with the *y*s, it is still possible both for the *x*s and for the *y*s to underlie the same Unity. We bring this distinctive feature of Unities out by means of the following assumptions.

First, we point out that it is possible for the *x*s to underlie *y* without composing *y* (and thus - taking into account D14 - we are denying D4 of section II above):

A24 for some *x*s, for some *y*, the *x*s underlie *y* but do not compose *y*

And similarly, denying a direct implication of D4, we assume:

A25 for some *xs*, for some *y*, for some *zs*, the *xs* underlie *y* and the *zs* underlie *y*, but the *xs* are not coextensive with the *zs*

We do assume, however, with regard to *all* the elements of a Unity, that they not only underlie it but also compose it:

A26 for all *xs*, for all *y*, if the *xs* are all the elements of *y* then the *xs* underlie *y* and the *xs* compose *y*

This also guarantees that assumption A5 of section II is fulfilled, for A26 implies that

A27 if the *xs* underlie *y* then for some *zs*, the *xs* are among the *zs* and the *zs* both compose and underlie *y*

Finally, it is useful to have a term by means of which to describe collections whose members underlie a Unity. We shall speak of such collections as *embodying* collections, and define as follows:

D15 x embodies $y =_{\text{def.}}$
for some *zs*, the *zs* are all the members of *x*,
and the *zs* underlie *y*

Our assumptions obviously entail that many distinct collections may embody one and the same Unity. This suggests an interesting perspective from which to compare the conception of a whole as a Unity with its conception as a sum. Conceiving of a whole as a sum, we imagine that the relation between a whole and its parts can be satisfactorily represented by means of a dyadic predicate '*Rxy*', satisfied in case a term designating some whole is substituted for '*x*' and a term designating any of its parts is substituted for '*y*'. Conceiving a whole as a Unity, we rather imagine the relation between a whole and its parts as represented by means of a different dyadic predicate, '*Sxy*', which is satisfied in case (if '*x*' is substituted by a term designating some whole) '*y*' is substituted by a term designating not any *part* of the whole, but rather any *embodying collection of parts* of the whole. I shall not pursue this suggestive perspective at present, however.

5. Loose Uses of '*is a proper part of*'

According to the theory as outlined so far, instances of the relation *is a proper part of* fall under two species: Some are instances of *is a proper sub-collection of*, and others instances of *is a pre-element of*. Thus we account for whole-part relations between collections and their sub-collections, and between Unities and their pre-elements. It is natural to wonder whether one should not account also for other apparent whole-part relations, as, for example, the relation that would seem to hold between a Unity and a collection of *some* (i.e. more than one) of its elements. Such a relation is not countenanced in our account, as developed so far.

Furthermore, as we have seen, different Unities can be simultaneously sums of precisely the same pre-elements. This means that one Unity can be at the same place at the same time as another Unity. Since two such Unities cannot share all their pre-elements (otherwise they would be identical according to A23), the following question arises: How should one describe the relations between a Unity u and some pre-element x of a superposed Unity v, where x is not a pre-element of u.

One way in which one might introduce such relations would be by developing the theory in a way which renders all these varied types of relations as further species of the relation *is a proper part of*. My attempts so far (which I do not record here), however, indicate that this would lead to a substantially more complicated theory, with few evident advantages.

An alternative approach, which is both simpler and possessed of an intrinsic plausibility, is to render such apparent instances of *is a proper part of* as not being *bona fide* instances of that generic relation. On this account, using 'part' to refer to such relations involves a loose usage, one which should be distinguished from the strict usage associated with the relation *is a proper sub-collection of* and *is a pre-element of*.

I now sketch in outline, how such an account can be provided. One type of case may be accounted for by the simple means of introducing the notion of an *indirect part*. On the principles of the theory expounded above, a molecule m which is a part of some brick b cannot be, strictly speaking, a part of a collection of bricks, one of whose members is b. Supposing the brick to be an individual, and therefore a Unity, the molecule is a pre-element of the brick, and therefore a part of the brick (in that it bears to the brick a relation associated with one of the two species of *is a proper part of* which we have distinguished). The brick, on the other hand, is a sub-collection of the collection, and therefore a part of the collection (bearing to the collection a relation associated with the other of the two species of *is a proper part of*).

However, the molecule is neither a pre-element of the collection (for multi-membered collections have no pre-elements), nor a sub-collection of the collection (for only bricks and collections of bricks are among the sub-collections of the collection). So the molecule is not a part of the collection. However, the molecule is a part (i.e. a pre-element) of a part (i.e. a sub-collection) of the collection of bricks. This, as I explained above, does not conflict with the transitivity of *is a part of* (recall the relations *is PS-shape to* and *is PS-colour to*). I now propose to describe the molecule as an *indirect* part of the collection, defining as follows:

D16 x is an indirect part of $y =_{\text{def.}}$
x is neither a sub-collection of y nor a pre-element of y
but for some z,
x is either a sub-collection of z or a pre-element of z and
z is either a sub-collection of y or a pre-element of y

Other problematic circumstances seem to be found only in cases where for some *x*s, several distinct entities, y, z, etc., are each a whole which corresponds to the *x*s. For example, if each of the *x*s is a Unity, then on the one hand there is y, the collection of the *x*s. On the other hand, there may be a Unity z, which is underlain by the *x*s; furthermore, there may be another Unity, w, which is underlain by elements which are in turn underlain respectively by elements from among the *x*s; etc.

In order to deal with such cases, I propose the following adaptation of the notion of superposition (taking into account our distinction between two species of the relation *is a part of*):[28]

D17 u is superposed with $v =_{\text{def.}}$
the *x*s compose u and the *x*s compose v
(either in the sense of 'compose' associated with *is a pre-element of* or in the sense associated with *is a sub-collection of*).

We may now introduce the notion of a *vicarious part*. For example, suppose Statue is a clay statue, and suppose Lump is the lump of clay which constitutes it (at time t). The theory of Unities is consistent with the

[28] See definition of 'superposition' in section III of Chapter 5.

claim that Lump is not identical to Statue (as well as with the claim that they are identical). Supposing they are not identical, the resources of the theory of Unities allow us to account for this state of affairs in various ways. An outline of one such possible account is as follows (this account is obviously over-simplistic, but it is only meant to illustrate the principle).

We assume Lump is a Unity, the elements of which are so-and-so many millions of clay particles. Call these particles 'the *ps*'. Statue, on the other hand, is also a Unity, with elements Arm1, Arm2, Leg1, Leg2, Torso, Neck, and Head. Each of these elements of Statue, in turn, has some of the *ps* as its elements.

Arm1, on this hypothesis, is a part of Statue (for it is an element of it). It is not a part of Lump, however, for since Lump is a Unity, its only parts are Lump itself and its pre-elements, which are the clay particles and *their* pre-elements. We may, however, describe Arm1 as a *vicarious* part of Lump. We explain this notion as follows:

D18 *x* is a vicarious part of $y =_{\text{def.}}$
x is neither a sub-collection of *y* nor a pre-element of *y*
but for some *z*, such that *y* is superposed with *z*,
x is either a sub-collection of *z* or a pre-element of *z*

Accounting for cases which do not fall under the rubric of *bona fide* instances of the relation *is a proper of* in this way, i.e. as instances of the relations *is an indirect part of* and *is a vicarious part of*, would seem to satisfy the intuitive pull towards viewing (for example) a molecule as a 'part' of a collection of bricks, and Statue's Arm1 as a 'part' of Lump. Strictly speaking, we should not describe these as parts; but loosely speaking we may, for the relations between the molecule and the collection, and between Arm1 and Statue, are in obvious ways derivable from, and associated with, relations which are strictly speaking relations between parts and wholes.

Appendix: Proofs of Selected Principles

Section II

Proof of C1:

1. the *x*s are parts of *y*
   [assumption]
2. the *z*s are among the *x*s
   [assumption]
   3. each of the *x*s is a part of *y*
   [from 1 and A1]
4. each of the *z*s is a part of *y*
   [from 2 and 3]
5. the *z*s are parts of *y*
   [from 4 and A1]
6. C1
   [from 1, 2 and 5]

Proof of C2:

1. the *x*s are parts of *y*
   [assumption]
2. *y* is a part of *u*
   [assumption]
3. each of the *x*s is a part of *y*
   [from 1 and A1]
4. each of the *x*s is a part of *u*
   [from 3 and 2]
5. the *x*s are parts of *u*
   [from 4 and A1]
6. C2
   [from 1, 2 and 5]

Section III

Proof of C1:

1. *y* is a member of *u* $=_{\text{def.}}$
   for any *x*s, if *u* is a collection of the *x*s then *y* is one of the *x*s
   [D2]
2. for any *y*s, for all *u*, for any *x*s,
   all of the *y*s are members of *u* (i.e. each of the *y*s is a member of *u*) iff if *u* is a collection of the *x*s then each of the *y*s is one of the *x*s
   [from 1]
3. for any *y*s, for all *u*, for any *x*s,
   all of the *y*s are members of *u* iff
   if *u* is a collection of the *x*s then the *y*s are among the *x*s
   [rewriting 2]
4. for any *y*s, for all *u*, for all *z*,
   only the *y*s are members of *u* iff if *z* is a member of *u* then *z* is one of the *y*s
   [analysis of 'only']
5. for all *z*, for all *u*, for any *y*s, for any *x*s, if *z* is a member of *u* then *z* is one of the *y*s iff if *u* is a collection of the *x*s then if *z* is one of the *x*s then *z* is one of the *y*s
   [from D2]
6. for all *z*, for all *u*, for any *y*s, for any *x*s, if *z* is a member of *u* then *z* is one of the *y*s iff if *u* is a collection of the *x*s then the *x*s are among the *y*s
   [from 5]
7. for any *y*s, for all *u*, for any *x*s,
   only the *y*s are members of *u* iff if *u* is a collection of the *x*s then the *x*s are among the *y*s
   [from 4 and 6]
8. C1
   [from 3 and 7]

Proof of C2:

1. for all *u*, if *u* is a collection of which all and only the *x*s are members, then *u* is a collection of the *x*s
   [from C1]
2. for all *u*, if *u* is a collection of which all and only the *x*s are members, then each of the *x*s is a Unity
   [from 1 and A5].

3. if *u* is a collection whose only member is *x*, then *x* is a Unity
[from 2]
4. C2
[from 3 and A7]

Proof of C3:

1. for all *u*, if *u* is a single-membered collection, then *u* is a Unity
   [from C2]
2. for all *u*, if *u* is a Unity then for some *y*, *y* is a collection of the *xs*, and the *xs* = *u*
   [from A6]
3. for all *y*, for any *xs*, if *y* is a collection of the *xs* then *y* is a collection of which all and only the *xs* are members
   [from C1]
4. or all *u*, if *u* is a Unity then for some *y*, *y* is a collection whose only member is *u*
   [from 2 and 3]
5. for all *u*, if *u* is a Unity then for some *y*, *y* is a collection whose only member is *u*, and *y* = *u*
   [from 4 and A7]
6. for all *u*, if *u* is a Unity then *u* is a single-membered collection
   [from 5]
7. C3
   [from 1 and 6]

Proof of A15:

1. *x* is a proper sub-collection of *y*
   [assumption]
2. *x* is a sub-collection of *y*
   [by 1 and D4]
3. for all *z*, *z* is a member of *x* only if *z* is a member of *y*
   [by 2 and D3]
4. for all *y*, for all *x*, if *y* is a proper sub-collection of *x* then for some *w*, *w* is a member of *x* but not of *y*
[by D4]
5. *y* is not a proper sub-collection of *x*
   [by 3 and 4]
6. A15
   [by 1 and 5]

Proof of A16:

1. *x* is a proper sub-collection of *y*
   [assumption]
2. *y* is a proper sub-collection of *z*
   [assumption]
3. *x* is a sub-collection of *y*
   [by 1 and D4]
4. *y* is a sub-collection of *z*
   [by 2 and D4]
5. for all *w*, *w* is a member of *x* only if it is a member of *y*
   [by 3 and D3]
6. for all *w*, *w* is a member of *y* only if it is a member of *z*
   [by 4 and D3]
7. for all *w*, *w* is a member of *x* only if it is a member of *z*
   [by 5 and 6]
8. *x* is a sub-collection of *z*
   [by 7 and D3]
9. for some *w*, *w* is a member of *z* but not of *y*
   [by 2 and D4]
10. for some *w*, *w* is a member of *z* but not of *x*
   [by 9 and 5]
11. *x* is a proper sub-collection of *z*
   [by 8 and 10]
12. A16
   [by 1, 2, and 11]

Proof of A17:

1. *x* collection-overlaps *y*$=_{def.}$
   for some *z*, *z* is a sub-collection of *x* and *z* is a sub-collection of *y*
   [definition]

2. *x* is collection-disjoint from *y* $=_{def.}$
   it is not the case that *x* collection-overlaps *y*
   [definition]
3. *x* is a proper sub-collection of *y*
   [assumption]
4. for some *z*, *z* is a member of *y* but not of *x*
   [by 3 and D4]
5. for all *z*, if *z* is a member of *y* then for some *v*, *v* is a collection

whose only member is *z*, and *v* is a sub-collection of *y*
[by A6, C3 and D3]

6. for all *v*, for all *z*, if *v* is a collection whose only member is *z*, and if *z* is not a member of *x*, then *v* is collection-disjoint from *x*
[by 1 and 2]
7. if *z* is a member of *y* but not of *x*, then for some *v*, *v* is a collection whose only member is *z*, and *v* is a sub-collection of *y*, and *v* is collection-disjoint from *x*
[by 5 and 6]
8. for some *v*, *v* is a sub-collection of *y* and *v* is collection-disjoint from *x*
[by 4 and 7]
9. A17
[by 3 and 8]

Proof of A18:

1. *y* is a collection-sum of the *xs* $=_{\text{def.}}$
for all z, z is a member of *y* if and only if for some *u*, *u* is one of the *xs* and *z* is a member of *u*
[definition]
2. the *xs* = $x_1, \ldots, x_n =_{\text{def.}}$
z is one of the *xs* iff *z* is identical to x_1 or to ... or to x_n
[definition]
3. the *xs* = $x_1, \ldots, x_n$
[assumption]
4. for $i = 1 \ldots n$, x_i has exactly $m(i)$ members, and
for j = 1 ... m (*i*), u_{ij} is the j-th member of collection x_i
[assumption]
5. the *zs* = $u_{11}, \ldots, u_{1m(1)}, \ldots, u_{n1}, \ldots, u_{nm(n)}$
[assumption]
6. each of the *zs* is a Unity
[by A5, 4 and 5]
7. for some *y*, *y* is a collection of the *zs*
[by A6, 6]
8. for any *xs*, for some *y*, *y* is a sum-collection of the *xs*
[by 1-5 and 7]
9. A18
[8]

Chapter 10
Further Elaborations and Applications

I - Perspectives on the Theory of Unities

1. Conditions of Identity

We are familiar with assumptions according to which the identities of comprising entities are bound up with the identities of entities they comprise. Such assumptions provide us with conditions for the identity of a comprising entity stated with reference to the identities of the comprised entities. These conditions may be stated, noting in each case in brackets the relation which establishes the identity dependence, as follows. In the case of classes, we have:

Identity (via *is a member of*) for Classes
for all *u*, for all *v*,
if *u* and *v* are classes, then $u = v$ if and only if
for all *x*, *x* is a member of *u* if and only if *x* is a member of *v*

In the case of classical sums, we have the following similar assumption:

Identity (via *is a part of*) for Classical Sums
for all *u*, for all *v*,
if *u* and *v* are classical sums, then $u = v$ if and only if
for all *x*, *x* is a part of *u* if and only *x* is a part of *v*

These assumptions express a feature of classes and classical sums known as their *extensionality*. Neoclassical mereological principles, by contrast with those governing classes and classical sums, do not imply the extensionality of the comprising entities they countenance. For with regard to neoclassical sums the identity of parts is only necessary, and not sufficient, for the identity of their sums:

Identity (via *is a part of*) for Neoclassical Sums
for all u, for all v,
if u and v are neoclassical sums, then $u = v$ only if
for all x, x is a part of u if and only if x is a part of v

It is by giving up the assumption that the identity of parts is sufficient for the identity of their sums, that neoclassical theory allows sums to be introduced which are superposed with one another.

A condition of identity for Unities may be stated which is analogous to the one applying to classes and classical sums (similarly contrasting with the condition which applies to neoclassical sums):

Identity (via *is an element of*) for Unities
for all u, for all v,
if u and v are Unities, then $u = v$ if and only if
for all x, x is an element of u if and only if x is an element of v

We see then that Unities share with classes and classical sums the feature of extensionality.

The relations *is a member of*, *is a part of*, and *is an element of* are relations, in the case of each type of comprising entity, between *each* of the comprised entities and the comprising entity. Identity determination between comprised and comprising entities runs, however, not only through these relations but also through relations which do *not* hold between each of the comprised entities and the comprising entities. Regarding wholes, 'make up' was the pre-theoretical term we used for such relations. In the case of classical and neoclassical mereology this was interpreted as the relation *compose*; in the theory of Unities, as the relation *underlie*; and in the case of set-theory as the relation (I have dubbed as) *form*. We have conditions of this type for classes and classical sums as follows:

Identity (via *form*) for Classes
for all u, for all v, for any xs, for any ys,
if u and v are classes, and the xs form u and the ys form v,
then $u = v$ if and only if
for all z, z is one of the xs if and only if z is one of the ys

Identity (via *compose*) for Classical Sums
for all *u*, for all *v*, for any *x*s, for any *y*s,
if *u* and *v* are classical sums, and the *x*s compose *u* and the *y*s compose *v*, then *u* = *v* if and only if
for all *z*, *z* overlaps one of the *x*s if and only if *z* overlaps one of the *y*s

In the case of neoclassical sums, there is a condition similar to that just stated for classical sums, the difference being that in the neoclassical case a conditional replaces the biconditional, as follows:

Identity (via *compose*) for Neoclassical Sums
for all *u*, for all *v*, for any *x*s, for any *y*s,
if *u* and *v* are neoclassical sums, and the *x*s compose *u* and the *y*s compose *v*, then *u* = *v* only if
for all *z*, *z* overlaps one of the *x*s if and only if *z* overlaps one of the *y*s

We see then that classes compare in this respect with sums (both classical and neoclassical) as follows: For any *x*s to determine the same class as do the *y*s, the *x*s have to be *identical* to the *y*s. By contrast, for any *x*s to determine the same sum (classical or neoclassical) as do the *y*s, the *x*s have to be *coextensive* with the *y*s.

In the case of Unities, the determining relation requires neither identity nor coextensivity. That is, for any *x*s to determine the same Unity as do the *y*s, the *x*s neither have to be identical to, nor coextensive with, the *y*s. In contrast with the case of *neoclassical* sums, however, the extensionality of Unities implies that *if* (to emphasize: 'if', not 'if and only if') the *x*s are severally identical to the *y*s then the Unities they underlie are identical:

Identity (via *underlie*) for Unities
for all *u*, for all *v*, for any *x*s, for any *y*s,
if *u* and *v* are Unities, and the *x*s underlie *u* and the *y*s underlie *v*,
then *u* = *v* if
for all *z*, *z* is one of the *x*s if and only if *z* is one of the *y*s

The contrast between the different ways in which the identity of comprised entities determines the identity of comprising entities, in connection with classes, classical (and neoclassical) sums, and Unities, suggests that we should distinguish these as three general types of comprising entity. In Part One it was explained that modern theorizing about comprising entities is characterised by a distinction between two basic types of comprising entity: That of distributive classes, and that of collective classes. Classes, as featuring in set-theory, are a development of the former type. Sums, as featuring in both classical and neoclassical mereology, are alternative developments of the latter type. The fundamental nature of the contrast between those two basic types of comprising entity is exhibited by the contrast between the identity-determining relations explained above.

Again, in the case of classes, the *x*s and the *y*s may form the same class only if the *x*s are identical to the *y*s. In the case of sums, the *x*s and the *y*s may compose the same sum only if the *x*s are coextensive with the *y*s. The fact that in the case of Unities, neither of these requirements apply, indicates that the notion of a Unity should be taken as that of a third basic type of comprising entity, alongside classes and sums.

2. Analogy with Other Relations

In order to grasp more clearly what the relation *underlie* involves, and what sort of identity dependence it establishes between elements of a Unity and the Unity they underlie, contrasting this dependence with those established by the relations *form* and *compose*, it will be helpful to consider a similar contrast which occurs in relations which are not those of parts to wholes. A rather close analogy to the contrast between *underlie* and *compose* can be found among relations between geometrical points and figures which they specify. In order to avoid complications arising from the nature of abstract entities, I shall consider dots on a blackboard instead of genuine geometrical points.

If three dots are marked on an otherwise clear blackboard, then there is one and only one circle which could be drawn through these dots (ignoring the fact that the dots have a finite size, and therefore that they do not specify in strict precision the course through which the circle will run). Let us think of this as an invisible circle which actually runs through the dots (perhaps it is only visible through infra-red sensitive goggles). There are infinitely many different (invisible) circles which run through any two of these dots, but only one which passes through all three of them. The three dots might thus be said to specify a single circle.

Now suppose not three but ten dots are marked on the blackboard, such that a single circle could be drawn through them all. Clearly, any three of the ten dots specify the same circle (which passes through them all). Moreover, any four dots, or indeed any n dots (where n is greater than 2) of those marked would serve the same purpose. A relation which may be called 'circle-specify' thus holds between any three or more of the dots and the circle that is drawn through them.

We can now see that the relation between the circle and ten dots which lie on it is analogous to the relation between a Unity and its elements. The relation *circle-specify*, as holding between dots and circles, is analogous to the relation *underlie*, as holding between elements and Unity. Just as several of the Unity's elements (not necessarily all) may underlie it, so several of the dots, (not necessarily all ten of them) may circle-specify the circle. And just as non-coextensive collections of elements may underlie the same Unity, so non-coextensive collections of dots may be such that the members of either of them circle-specify the same circle.

It must be emphasized, though, that the relation *circle-specify*, as holding between dots and circles, is only *analogous* to the relation *underlie*, and should by no means be assumed to be an *instance* of it. The relation *circle-specify*, indeed, is not a relation between parts and a whole. The dots are made of chalk. The circle, we have assumed, is invisible, and has no parts made of chalk.[1] The relation *underlie*, by contrast, is a relation between parts and a whole which they make up.

Now, consider in stead the same ten dots as specifying not circles, but rather polygons whose vertices coincide with dots. Taking any three of the dots, there is clearly one and only one polygon all of whose vertices coincide with those dots, and this polygon, of course, is a triangle. The three dots may then be said to specify that triangle. Let us call the relation of specification between dots and polygons, thus conceived, 'polygon-specify'.

It is clear that if the *x*s and the *y*s are any dots among the ten dots on the blackboard, the *x*s polygon-specify the same figure as do the *y*s if and only if the *x*s are identical to the *y*s.[2] Thus the relation *polygon-specify*, as holding between dots and polygons, is analogous to the relation *form*, as holding between members of a class and the class itself.

To construct an analogy with *compose*, as holding between parts and classical sums, consider not the dots individually, but rather collections of

---

[1] However, even if both the dots and the circle were made of chalk, the ten dots would obviously not make up the circle, and so the circle would not be a whole which corresponds to the ten dots.

[2] Again, we say the *x*s are identical to the *y*s in case for all *z*, *z* is among the *x*s if and only if *z* is among the *y*s. I have sometimes used the expression 'severally identical' above to express such identity claims holding between the *x*s and the *y*s.

dots,[3] as specifying polygons. Thus suppose there are pairs, triplets, etc. of dots. If, for example, *x* is a pair of dots, and *y* is a triplet of dots, then *x* and *y* can be taken to specify a polygon as follows: There is one, and only one polygon, such that all and only those dots that are members either of *x* or of *y* coincide (respectively) with vertices of the polygon. The relation between collections of dots and the polygons they specify in this way may be called 'co-polygon-specify'.

Inspection reveals that if the *x*s are (one or more) collections of dots, and the *y*s are (one or more) collections of dots, then the *x*s co-polygon-specify the same figure as the *y*s if and only if the *x*s are coextensive with the *y*s. The relation *co-polygon-specify*, as holding between collections of dots and the polygons they specify, is thus clearly analogous to the relation *compose* as featuring between parts and their classical sum.

Finally, if one wishes to construct an analogy with *compose*, as featuring in connection with neoclassical sums, one may modify the last example by assuming that each collection of dots might specify not one, but two or more invisible polygons which are superposed with one another (we might take each such polygon to be visible through different types of goggles - infra-red sensitive goggles, ultra-violet sensitive ones, etc.).

Again, it should be emphasized that each of the relations presented between dots and figures are to be taken as analogous to, rather than as instances of, the corresponding comprising relations. Four ways have been noted - according to which dots, or collections of dots, on the blackboard can be construed as specifying figures. In each case, the identities of the dots, or of the collections of dots, are seen to determine the identity of a figure in a particular way.

Focusing on the conditions which comprised entities necessarily meet if the entity which comprises them is to be the same, the cases are seen to fall into three types, corresponding to *circle-specify* (as holding between dots and circles), *polygon-specify* (as holding between dots and polygons), and *co-polygon-specify* (as holding between collections of dots and polygons). These relations exhibit patterns of identity determination precisely analogous to those exhibited by *underlie* (between elements and their Unities), *form* (between members and their classes), and *compose* (between parts and their sums - classical or neoclassical) respectively.

---

[3] I use 'collection' here in the more general sense presented in Chapter 3, rather than in the narrower sense pertaining to the theory of Unities.

3. The Character of Unities

Let us return to a point raised briefly in Chapter 9, section II. Consider again the house which has been built out of bricks, $b_1 \ldots b_n$ (we will abbreviate '$b_1 \ldots b_n$' as 'the *b*s'). Suppose beside the house there has been left a pile of unused bricks. If we describe the house as a sum composed by the *b*s, then the notion of a sum guarantees that while the *b*s are parts of the house, the bricks in the pile are not parts of the house. If the *b*s+*c*s are all the bricks of the house *and* the bricks in the pile, then if there is a sum of the *b*s+*c*s, it is clear that this sum is not identical to the sum of the *b*s. For there are entities which overlap the sum of the *b*s+*c*s which do not overlap the sum of the *b*s.

However, if the house is described as a Unity underlain by the *b*s, then there is nothing in the assumptions so far made regarding Unities which would preclude the possibility that not only the *b*s but also the bricks in the pile are elements of the house. It might be thought that if that is so, the notion of a Unity and of the relation *is an element of* cannot appropriately be considered as providing an account of our notions of whole and part. For it might be thought that if the identity of a whole is determined (by identifying parts which make it up) then it must be possible to determine on this basis regarding any entity whether or not it is a part of the whole. In conflict with this, the theory of Unities seems to allow us to determine the identity of the house without enabling us to deduce from this that the bricks lying beside it are not parts of it.

Such an objection, however, is based on a misunderstanding. True, the assumption that the house is composed of the *b*s implies that the house is not composed of the *b*s+*c*s, whereas the assumption that the house is underlain by the *b*s does not imply that the house is not underlain by the *b*s+*c*s. However, in either case the respective assumptions derive from our understanding of the concept of a house. It is on the basis of this understanding that one is lead to claim either that the house is composed by the *b*s or that it is underlain by the *b*s. Furthermore, it is precisely on the basis of this understanding that one is led to claim, either that the house is not composed of the *b*s+*c*s, or that it is not underlain by the *b*s+*c*s.

The only difference between the cases is that, whereas in case we conceive of the house as a sum we may arrive at the belief that the house is not composed of the *b*s+*c*s not only directly on the basis of the concept, but also indirectly, on the basis of the assumption that the house is composed of the *b*s (where this in turn is derived directly from the concept), we are not given such an indirect inferential path in case we conceive of the house as a Unity. This, however, in no way weakens our justification for thinking that the house is not underlain by the *b*s+*c*s. The claim - given the assumption that a house is a sum - that the house is not composed by the *b*s+*c*s, rests on precisely the same foundation as the claim - given the assumption that a house is a Unity - that the house is not underlain by the *b*s+*c*s, and so in this respect both claims are justifiable to the same degree.

Far from being a disadvantage of the theory of Unities, the lack of entailment between the claim that the *b*s underlie a Unity and the claim that the *b*s+*c*s do not underlie a Unity is a feature of the theory which exhibits striking advantages in a variety of cases. I shall give two examples.

Vagueness in Sortal Concepts

It is commonly observed that many of our concepts for types of entity are vague, in the sense that the concept seems not to determine clearly with regard to some things, whether or not they are parts of the entity. For example, the concept of a battlement (in the sense of certain part of a castle)[4] would often seem not to determine whether a certain stone which is a part of the castle is or is not also a part of the battlement. One is likely to be unsure where the battlement ends and the rest of the castle begins.

If a battlement is considered to be a sum of stones, then the vagueness in the concept of a battlement reads as a vagueness as to which stones compose the battlement. The concept does not allow us to say determinately with regard to *any* stones, that they compose the battlement. But this in turn becomes a vagueness as to the *identity* of the battlement.

For suppose that the *x*s and the *y*s are stones, where some of the *y*s are not among the *x*s, and suppose sums exist of both the *x*s and the *y*s, so that a sum of the *x*s is distinct from a sum of the *y*s. Suppose that the difference between the two sums is clearly visible; the *y*s include ten extra stones constructed to the left, adjacent to the *x*s. Suppose we are unproblematically able to identify these two sums. However, since the vagueness in the concept makes us doubt whether the battlement is

[4] *The Concise Oxford Dictionary of Current English* gives the following definition of 'battlement': 'Indented parapet ...; parapet & enclosed roof.'

identical to the sum of the *x*s or to the sum of the *y*s, then since the sums are distinct we are in doubt about the identity of the battlement. It turns out then that on the conception of a battlement as a sum, the vagueness in the concept of a battlement becomes a vagueness as to which entity (which of the two distinct sums) the battlement is identical to.

Now consider the alternative conception of a battlement as a Unity. Of course, such a change in conception does not make the vagueness in the concept of a battlement disappear. However, it can readily be seen that now this vagueness does not lead to vagueness regarding the identity of the battlement. Rather, the vagueness is confined merely to the question regarding some of the stones whether they are elements of the battlement or not. Conceiving the battlement as a Unity, we are admittedly still unclear about the status of some stones as elements of the battlement, but we are not unclear about the identity of the battlement: We are not unclear, as before, as to whether the battlement is one or the other of two distinct, existent wholes. A vagueness affects the identity of the complete embodying collection of stones, but not the identity of the embodied Unity itself .

The reason for this is that when in doubt as to some stones whether they are among the elements of the Unity, one may still be clear with regard to *some* stones that they underlie the battlement. If the *x*s are such stones, where each of them is *determinately* (rather than vaguely) an element of the battlement, and the *x*s underlie the battlement, then the identity of the battlement is fixed as a Unity underlain by the *x*s. Whether or not for some *y*s (where not all of the *y*s are among the *x*s) the *y*s *also* underlie the battlement in no way involves a recasting of the battlement's identity. The remaining vagueness pertains only to the question whether the additional stones are to count also as elements of the same Unity.

Pre-theoretically we do not seem to be in doubt as regards the identity of the battlement, although we may well be in doubt as to whether a certain stone is one of its parts. We do not think that there are two, or perhaps a hundred different entities, largely overlapping one another, any of which might be the battlement. If someone points to a battlement belonging to the castle over there, I am not inclined to ask her which of the many sums that I am able to identify she takes to be the battlement. It is clear that the conception of wholes as Unities is better suited to capturing common sense identificational practice in this respect.

An account along the lines presented here regarding the case of the battlement applies as a general treatment of the omnipresent cases in which Unger's so-called 'problem of the many' arises.[5] Reflection on the received scientific portrayal of macroscopic concrete entities as constituted by swarms of micro-particles leads to viewing the former much like dust-clouds or puffs of smoke, entities which do not display sharp boundaries. At the fringes of any concrete macro-entity, according to this portrayal, there are micro-entities with regard to which it seems to be a vague matter whether they are, or are not, parts of the macro-entity.

Suppose an entity *s* is constituted by such a swarm of particles, where the *p*s are particles that are definitely parts of the swarm, and the *q*s, that is $q_1, \ldots, q_n$, are particles, whose status as parts of the swarm is indeterminate.[6] Suppose the *p*s and q_i have a sum s_i (for all $i = 1, \ldots, n$), and suppose all the s_i have equal claim to being the entity *s*, and furthermore that no sum of particles has greater claim to being *s* than any of the s_i. If we take *s* to be a sum of particles, then there would seem to be no fact of the matter as to *which* of the sums of particles (i.e. which of the s_i) is identical to *s*. Thus the identity of *s* is rendered vague.

The problem with the conception of wholes as sums in this context is that it implies that for *s* to be a whole which corresponds to the *p*s and q_i is incompatible with its being a whole which corresponds to the *p*s and q_j (where $i \neq j$). If we assume that only one of these wholes exist, then there seems to be no hope of explaining *why* this one rather than the other exists. For there is no difference between the two cases which would reflect on the abilities of the respective particles - the *p*s and q_i, on the one hand, and the *p*s and q_j, on the other - to have a sum. If alternatively we assume that both wholes exist, we are faced with the need to decide which of them is *s*, leading us to view the identity of *s* as vague.

According to the conception of wholes as Unities, by contrast, the two claims (in the last paragraph) *are* compatible with one another. If *s* is a Unity, it might be both a whole which corresponds to the *p*s and q_i and a whole which corresponds to the *p*s and q_j. Choosing one of these options does not compel us to reject the other. We are thus led to no indecision regarding the identity of *s*, and need not consider the identity of *s* as vague.

David Lewis surveys a number of responses to the problem of the many, which share the (unstated) conception of wholes as sums.[7]

---

[5] See Unger 1981.

[6] I am not assuming that the *p*s are *all* the particles which are determinately parts of *s*, nor that the *q*s are *all* the particles which are indeterminately parts of *s*.

[7] See Lewis 1993.

According to one proposal (due to Lowe), each of the *n* sums is an entity which constitutes *s*, but constitution, on this view, is distinct from identity, and there is no contradiction in assuming that many distinct entities may constitute a single entity. According to a second proposal, *s* is assumed to be a vague entity, and each of the *n* sums is taken to be a 'precisification' of *s*. A third proposal (based on van Fraassen's method of supervaluations), like the second one, concedes that there is no fact of the matter as to which of the *n* sums is identical to *s*, and yet, by contrast with the second proposal, argues that the claim that one of those sums is identical to *s* is true nevertheless, if this claim is interpreted according to the supervaluational conception of truth. According to a fourth proposal (following Geach), the *n* sums are different particle-sums, but the same swarm; identity is taken to be relative to sortal term. And finally, on a fifth proposal (contributed by Lewis himself, on the basis of Armstrong's notion of partial identity), identity admits of degrees, and the *n* sums are 'almost' identical to one another, and *s* is as it were almost one entity (and almost of determinate identity).

Limitations of space do not allow me to discuss these responses in detail. I shall only add the following brief comment. These views have it in common that they attempt to account for the identity of an entity with indeterminate boundaries (e.g. a cloud) in terms of the identities of entities with determinate boundaries (e.g. precisely delimited swarms, or sum, of particles). The conception of wholes as sums compels them to seek such an account. For it takes wholes to be entities which are *coextensively determined*, so that the *x*s cannot make up the same *u* as the *y*s unless the *x*s are coextensive with the *y*s.

The result is that the identity of a whole becomes inseparably bound up with the identity of its boundary. But the case (of the cloud, as of other macroscopic entities) does not allow them to locate a *unique* precise entity, so they invariably end up explaining the identity of the imprecisely delimited entity, which on the face of it is unique, in terms of many almost perfectly overlapping precisely delimited entities. The conception of wholes as Unities does not bind the identity of a whole to the identity of its boundary (unless the whole happens to have a clearly identified boundary), and so is not in the predicament of these theorists.[8]

---

8 Interestingly, the constraints of the conception of wholes as sums affect even Lowe's account, which seems to come closest among currently existing views to breaking loose with it. His account still presupposes that wholes are fundamentally sums, but allows that they need not be determinately one sum or another. If the *x*s are not coextensive with the

Partly Perceived Entities

We often perceive entities which are partly hidden from view. For example, suppose Tibbles the cat is sitting in the doorway, in such a way that its tail cannot be seen. Now, if we are unfamiliar with the cat, we might entertain a doubt as to whether indeed it has a tail. Suppose we consider the cat as having the following parts: Head, Neck, Torso, Leg1, Leg2, Leg3, Leg 4, and possibly Tail.

If we take the cat to be a sum (whether classical or neoclassical) of its parts, then the following difficulty appears. It would seem plausible to assume that the visible parts of the cat (i.e. the first seven in the list above) have a sum, whether or not Tail exists. It is not clear whether an assumption to the contrary, namely, that a cat which has a tail does not have a sum which includes everything but the tail, could be satisfactorily motivated.[9] Furthermore, this assumption would imply that we are not in a position to judge whether the visible parts of the cat do or do not have a sum, unless we know whether or not the cat has other parts, outside the range of our vision. This would mean that perceptual experience is usually unable to justify the claim that a whole made up of certain parts exist, assuming that whole is a sum, for we are usually not in a position to see enough of the parts (usually you see only one side of the cat you are looking at; and in any case, its internal organs are hidden from view).

If we assume, then, that the visible parts of the cat have a sum, whether or not Tail exists, then it would seem that we are in a position to see and to identify this sum. However, we are not in a position to say whether what we see, when we see the sum, is identical to the cat or not. For if Tail exists, then the cat is not a sum of its visible parts, but rather a sum of those parts plus Tail. This means that we are uncertain about the identity of the cat - we cannot say whether it is or is not identical to something which we see, namely, the sum of its visible parts. On the conception of

ys, a whole *u* may be *indeterminately* a sum of both. Whether he is right to assume that this avoids rendering the identity of the whole vague, I shall not go into. In any case it is understandable that Lewis should complain about Lowe's account, that it transfers the original problem from a problem about constituted concrete entities to a problem about constituting concrete entities. See Lewis 1993, 25-6; Lowe 1982.

[9] Van Inwagen makes such an assumption in van Inwagen 1981. Simons offers some criticism of this view in Simons 1987, 121. I consider the view in more detail in section III below.

wholes as sums, doubt as to whether an entity has parts which are hidden from view becomes doubt regarding the identity of that entity.

Again, this result is at odds with common sense. We are often in doubt as to whether an entity which we see has parts which we cannot see. This, however, does not lead us to doubt whether we have succeeded in identifying the entity. If we turn to the conception of wholes as Unities, we find that here too, as in cases associated with vague sortal concepts, this conception contrasts with the former one as being in harmony with common sense.

For if we take the parts of the cat to be elements, and the cat to be a Unity, then we may claim that the visible elements of the cat underlie the cat, whether or not the cat is also underlain by elements some of which are not visible. Thus it is both the case that the visible elements of Tibbles underlie Tibbles, and (if Tail exists) that the visible elements *plus* Tail underlie Tibbles. The fact that one of two distinct collections of elements embodies a Unity is consistent with the other collection embodying the same Unity. Therefore doubt as to whether the cat has a tail or not does not lead, on the conception of a cat as a Unity, to doubt regarding the identity of the cat.

4. The Analogy between Elements and Properties

The peculiar type of determinative relation which elements bear to the Unity they underlie has a particularly interesting analogue in the relation between properties and a concrete entity in which they inhere. Noting this analogy makes us further realise that the formal features of the relation *underlie* are not at all a strange special case in the realm of determinative relations. It would take us far beyond the scope of the present work to explore the issues raised by this analogy in any detail. However, relying on some simplifying assumptions (assumptions which are not implausible but which will not be argued for here), the basic point may be sketched briefly. Let us assume

(1) 'Property' can be taken in a broad sense, so that whatever is designated by any monadic predicate (of the lowest logical type) which *prima facie* designates either a non-relational or a relational property, is considered to be a property.[10]

[10] On a view which takes there to be an ontological category of properties, whether conceived as universals or as tropes, (1) is not very plausible, if it is assumed that

Furthermore, let us assume

(2) If *x* is an individual belonging to a certain broad class of individuals, *x* may be identified (in the actual world) with reference to several properties, such that *x* and only *x* has all those properties.[11]

Thus, for example, it is reasonable to assume that if Socrates occupied a region of space *s* (somewhere in Athens) at time *t* (some time in 400BC), then Socrates can be identified as the unique individual which has both of the properties *occupies region* s *at time* t and *is a man*. Let us say that if the *F*s are properties with reference to which *x* can be identified, then the class whose members are (all and only) the *F*s is an *identifying class* of *x*. Let us also say in such a case that the *F*s *determine the identity of x*. And finally, let us assume

(3) If *C* is an identifying class of properties of *x*, and if *F* is a property of *x* which does not belong to *C*, then the class whose members are *F* and the members of *C* is an identifying class of *x*.

Thus, for example, assuming the class whose members are the properties *is an Athenian philosopher* and *is executed in 399BC* is an identifying class of properties of Socrates, and assuming that *is snub-nosed* is a property of Socrates which does not belong to the latter class, we have it that the class whose members are *an Athenian philosopher*, *is executed in 399BC* and *is snub-nosed* is an identifying class of properties of Socrates.

On these assumptions, a clear analogy can be drawn between the relation *determine the identity of*, as holding between properties and a

---

'property' is to apply only to members of this ontological category. Armstrong convincingly criticises semantic arguments for the existence of universals, and urges that the use of a predicate in true statements provides poor evidence for the existence of a universal which is designated by the predicate (see e.g. Armstrong 1978, vol.1, 64). However, if in cases where the predicate does not designate such a real existent we may assume that the predicate designates a logical construction out of such real existents, we may take 'property' in a broad sense as covering not only the real existents but also logical constructions of this sort.

[11] The assumption that individuals have essential properties, such that those properties may serve to identify individuals in any possible world in which they exist is controversial (see Kripke 1972, Lecture I). In (2), however, it is only assumed that individuals have properties, not necessarily essential ones, which may serve to identify them in the *actual* world. This does not seem to be controversial.

concrete entity, and the relation *underlie*, as holding between elements and a Unity.

Just as some of the elements of a Unity, typically not all, may underlie the Unity, so some of the properties of a concrete entity, typically not all, may determine the identity of the entity. And just as both the *x*s and the *y*s may underlie the same Unity, even if they are not coextensive with one another, so the *F*s may determine the identity of an entity and the *G*s may determine the identity of the same entity, even if the *F*s are not coextensive with the *G*s.

We noted above that the conception of individuals as Unities allows us to provide a simple account for the phenomenon whereby we may identify partly hidden individuals. We are, of course, able to identify individuals perceptually, even if we perceive very few of the individual's (perceptible) properties. There is a clear analogy between the perceptual identification of an individual notwithstanding an only partial perception of its properties and the perceptual identification of an individual notwithstanding an only partial perception of its parts. Taking individuals to be underlain by their parts (thus taking the individuals to be Unities and their parts to be elements), and noting the similarities between *underlie* and *determine the identity of* (where the latter holds between some of an individual's properties and that individual), helps us account for this remarkable analogy.

II - Applying a Theory of Unities

1. Sketch of a Programme

I have already noted that I take the intended interpretation of the theory of Unities as a theory of concrete individuals (in what follows we speak of concrete individuals simply as individuals). A theory of Unities is developed, therefore, with the principal aim of applying this theory to the study of individuals. I propose as the first step, in developing the theory of Unities in application to individuals, the following simple and natural one:

PA1 for all u, if u is a individual, then u is a Unity

(where 'PA' stands for 'Principle of Application'). The theory of Unities is meant to provide a conceptual tool for accounting for various aspects of the nature of individuals. On the assumption that some concrete entity is an individual, it will be treated in terms of the theory as a Unity. It should not be expected, however, that such a theory can provide conclusive criteria on the basis of which we might decide whether some particular entity is an individual, or whether some particular type of entity is a type of individual. Nevertheless, the theory might offer us a range of considerations relevant to deciding on such questions.

For example, if considerations independent of the theory lead us to the view that a car is (collectively) identical to its proper parts (wheels, motor, body, etc.), then since a Unity is not identical to its elements, the proper parts of the car cannot be related to the car as elements to a Unity. Assuming that we take individuals to be Unities, this will lead us to the conclusion that the car is not an individual. However, unless such considerations external to the theory of Unities itself are brought to bear, the theory will be silent on whether the car is and individual or not.

A useful, fully developed theory of Unities can only be elaborated in conjunction with a detailed study of individuals, and in particular, a study of aspects of individuals which pertain to the relations they exhibit between parts and wholes. This is not the task of the present work. Nevertheless, I do want to suggest strongly that the theory of Unities is capable of fruitful application in the study of individuals, and that it can be developed as required for such a theoretical purpose. To do this, I propose what seem to me to be intuitively appealing principles which establish initial steps in the application of a theory of Unities. By doing so, however, I do not wish to imply that I take these principles to be essential, and that no alternative first principles of application might be subscribed to. With this in mind, let us look at several additional first principles guiding the way in which the theory can be applied and further elaborated.

Suppose u is an individual. Of course, this individual is a whole. According to PA1 u is a Unity. If it has proper parts which are themselves individuals, it follows that these proper parts are also Unities, and so according to D12 above they are pre-elements of u. That is,

PA2 for all *u* and for all *x*, if *u* is an individual and *x* is an individual, and *x* is a proper part of *u*, then *x* is a pre-element of *u* (that is, *x* is either an element of *u* or a proper pre-element of *u*)

For example, suppose a certain man *m* is an individual. And suppose a certain heart *h* is an individual, and is a proper part of *m*. According to principles PA1 and PA2 it follows that *h* is a pre-element of the Unity *m*. Similarly, if a certain cell *c* is an individual, and *c* is a proper part of *m*, then *c* too is a pre-element of Unity *m*. Furthermore, in view of the transitivity of *is a pre-element of*, *c* is also a pre-element of Unity *h*.

To help us determine whether or not a pre-element of an individual is an element of that individual (rather than a proper pre-element of the individual), we note that the intransitivity of *is an element of* implies that an element of an element of *u* cannot itself be an element of *u*. We have therefore

PA3 for all *x*, for all *u*,
if *x* is a pre-element of *u* then *x* is an element of *u* if and only if *x* is an immediate proper part of *u*

Applying PA3 to our example of a man and his proper parts, we conclude that the cell *c* is not an element of *m* (and so it is a proper pre-element of *m*). Furthermore, if we assume that no individual exists which is a proper part of the man and of which the heart is a proper part, then we conclude on the basis of PA4 that the heart *h* is an element of the man *m*. On the assumptions made so far, however, the theory of Unities is silent as to whether there is an individual - say the man's torso - which is a proper part of the man and of which the heart is a proper part.

Now, suppose we have identified several parts of an individual as being all its elements. The question we now face is, which of those elements are such as to underlie the Unity (or, to put the question in different terms, which collections of elements are collections which embody the Unity). An obvious principle of some relevance to this question is the following:

PA4 for any *x*s, for all *u*,
if the *x*s are all the elements of *u*, then the *x*s underlie *u*

It may now be concluded on the basis of PA4 that if we identify a great number of cells, the *c*s, as all the elements of heart *h*, then the *c*s underlie *h*. PA4, however, tells us nothing as to whether, for example, all those cells but one also underlie the heart, or whether a certain collection of cells to which 98% of the heart's cells belong is a collection which embodies the heart.

To deal with the question whether some incomplete collection of elements of a Unity embodies the Unity (or equivalently whether the members of such a collection underlie the Unity), a much more powerful principle than PA4 is required.[12] One promising idea in this connection is to interpret conditions under which a Unity survives as evidence for determining which are the embodying collections of elements. More specifically, if a Unity survives the loss of an element at *t*, then the remaining elements underlie the Unity after *t* (according to PA4). The proposal then is as follows: If the remaining elements are in much the same condition (as regards their properties and relations to one another) after *t* as before *t*, then they can be assumed to have been underlying the Unity *before t*, and not only after.

Let us say that when the condition of certain elements, the *x*s, does not change in any way pertinent to their capacity to underlie a Unity, those elements are *not modified* (this introduces a technical sense of 'modify', of course). The following principle is supposed, therefore:[13]

PA5 for any *x*s, for any *y*s, for all *u*, for all *t*, for all *t*'
if *t*' is later than *t*, and each of the *x*s is distinct from each of the *y*s, then if
(1) the *x*s + the *y*s underlie *u* at *t*,
(2) the *x*s underlie *u* at *t*', and
(3) the *x*s are not modified between *t* and *t*',
then the *x*s underlie *u* at *t*

Suppose for example that a heart has *n* cells as its parts at *t*. If it would survive the destruction (without replacement) of any one of those cells between *t* and *t*', without its other cells being modified between *t* and *t*',

[12] I say that a collection of elements of *u* is 'incomplete' if and only if some element of *u* is not a member of the collection. Otherwise the collection is said to be 'complete'.

[13] the expression 'the *x*s + the *y*s' is used in formulating PA5 in the following sense: the *x*s + the *y*s underlie *u* if and only if for some *z*s, each of the *z*s is either one of the *x*s or one of the *y*s, and each of the *x*s and each of the *y*s is one of the *z*s, and the *z*s underlie *u*.

then according to PA5 the heart is underlain already at *t* by any *n* -1 of its cells.

I shall not enter a discussion of various ways in which PA5 might be altered or refined (a refinement which immediately suggests itself, for example, would be to replace (3) by a more flexible condition). Note that the principle even as it stands has the advantage of appealing to features (of individuals) about which we are amply informed by experience, and which are at least to some significant degree susceptible of empirical investigation.

For example, it is known that trees generally speaking can survive the loss of several of their principal branches. On the assumption that principal branches are among the elements of a tree (perhaps minor branches are elements of principal branches, and thus proper pre-elements of the tree), it may be concluded directly, by means of PA5, that if the *x*s are all the tree's elements except for one main branch, then the *x*s underlie the tree. And were we not sufficiently informed about this by experience, a variety of scientific procedures suggest themselves which would supplement our empirical knowledge appropriately.

In conjunction with the detailed study of the way the notion of a Unity can be applied in accounting for the nature of individuals, it is to be expected that generalizations will be formed regarding fundamental constraints on conditions under which individuals may be assumed to underlie a Unity. Such generalizations are developed on the basis of our understanding of the sortal concepts under which individuals of various types fall, together with scientific experience regarding such individuals. These constraints can then be formulated as the postulates of a more developed theory.

I shall now give examples of several such general constraints which suggest themselves in the course of initial experimentation with the conception of individuals as Unities. It has to be emphasized that the list that follows is a tentative one, and is subject to all manner of revision on further consideration. It is chiefly intended to suggest the *type* of constraints that one might want to integrate into the theory of Unities.

It seems reasonable to assume, for example, that if each of two distinct collections of cells embodies a heart, then those collections overlap one another: there are cells which are members of both collections (the collection of shared elements may be called the 'intersection' of the two collections). We have then

Weak Sharing
for any *xs*, for any *ys*, for all *u*,
the *xs* underlie *u* and the *ys* underlie *u* only if for some *zs*, the *zs* are both among the *xs* and among the *ys*

A stronger assumption would be that the intersection of any two embodying collections is itself an embodying collection. This could be formulated as follows:

Strong Sharing
for any *xs*, for any *ys*, for all *u*,
the *xs* underlie *u* and the *ys* underlie *u* only if for some *zs*, the *zs* are both among the *xs* and among the *ys*, and the *zs* underlie *u*

Suppose a certain castle is a Unity, underlain by several elements, among which are the castle's north wing and its south wing. Suppose the two wings do not overlap one another. Suppose the castle would survive if every one of its elements were destroyed with the exception of one of the wings. Then on PA5 either of the wings underlies the castle. If one wishes to maintain these assumptions about the survival of the castle, then one will need to reject both Sharing principles. Alternatively, one might accept either of these principles, and revise accordingly one's assumptions regarding the castle.

One may proceed to introduce constraints which involve not only the relations between any two embodying collections, but also relations between many or all embodying collections. First, it seems plausible to suggest that there is a core collection of elements which is a part of any embodying collection of the Unity:

Weak Minimal Collection
for all *u*, if *u* is a Unity, then for some *xs*, for any *ys*,
if the *ys* underlie *u*, then the *xs* are among the *ys*

Moreover, it might be suggested that such a core collection would itself be an embodying collection:

Strong Minimal Collection
for all *u*, if *u* is a Unity, then for some *xs*, for any *ys*,
if the *ys* underlie *u*, then the *xs* are among the *ys*,

and the *x*s underlie *u*

It is tempting to identify the notion of a minimal embodying collection with the notion of a collection of *essential* parts. An essential part is a part on whose existence (and on whose appropriate relations to other parts) the existence of the whole depends. Clearly if a whole has essential parts, and the whole is a Unity, then the essential parts must either be members of any embodying collection, or they must be pre-elements of members of any embodying collection.[14]

Complementing constraints regarding minimal embodying collections, we have a constraint regarding a maximal embodying collection. Assumption A27 above implies the following principle:

Strong Maximal Collection
for all *u*, if *u* is a Unity, then for some *x*s, for any *y*s, if the *y*s underlie *u* then the *y*s are among the *x*s and the *x*s underlie *u*

The observation that boundaries between individuals are often sharp suggests that the following principle has some plausibility:

Non-overlap of Unities
for all *u*, for all *v*, if *u* is a Unity and *v* is a Unity, and $u \neq v$, then for all *x*, *x* is not an element of both *u* and *v*

It is perhaps not out of place to note that a variety of constraining principles follow from the basic assumptions A1-A27 and D1-D18. For example, Supplementation of Proper Parts has been counted among the basic mereological principles with which the theory of Unities complies. In a comment regarding assumption A11 (see Chapter 9, section III), doubts were raised as to whether this principle should indeed be asserted. A consequence of this assumption is the following somewhat artificial, though not intolerable, constraint:

[14] More sophisticated versions of Strong Minimal Collection could be tailored to distinguish between different conceptions of the dependence involved between the whole and the parts. A human organism must, it would seem, have *a* heart, but *which* heart perhaps does not matter to the organism's identity (see Simons 1987, 294ff. on the distinction between generic and rigid dependence). Such distinctions would involve making explicit the temporal modification of 'underlie', and specifying constraints involving relations between collections which embody the Unity at different times.

Plurality of Underliers
for all *u*, for any *x*s,
if the *x*s underlie *u*, then the *x*s are at least 2 in number

Another example is the following principle, which is a consequence of the intransitivity of *is an element of*, and which suggests a hierarchical mereological conception of individuals:

Separation of Levels
for all *x*, for all *y*, if *x* is an element of *y*, then *x* and *y* cannot be elements of one and the same Unity

A great wealth of additional constraints suggest themselves in connection with processes involving the flux of parts of individuals. Such constraints may state, among other things, conditions for the survival of Unities in the face of mereological variation. For example, the following seems to be a plausible principle:

Dispensable Elements
for all *u*, for all *x*, for all *t*
if *x* is one of the elements of *u* at *t*
and *x* is not an element of *u* immediately after *t*, then
u exists immediately after *t* only if *x* is not a member of *every* embodying collection of *u* at *t*

The latter principle may be developed into a necessary condition for survival over time through a series of mereological changes, as follows:

Continuity
for any *x*s, for all *u*, for any *y*s, for all *v*, for all *t* and *t'*
if the *x*s underlie *u* at *t*, and *t* is earlier than *t'*, and the *y*s underlie *v* at *t'*, then *u* is identical to *v* only if
for all *z*, if *z* is one of the *x*s but not one of the *y*s, then
z had ceased to be an element of *u* under circumstances complying with the principle of Dispensable Elements

Just as much as principles may be sought generalizing conditions under which loss of parts may occur, so we should look into principles associated with *gain* of parts. I shall not elaborate such principles, as well as others which suggest themselves on reflection. What has been said should suffice to indicate how one might approach an explanatory programme utilising the notion of a Unity.

2. Dissolution of Neoclassical Problems

In Chapter 7 we looked at a number of fundamental difficulties that are involved in some of the more important types of neoclassical sum. A brief consideration suffices to show that these difficulties do not burden the theory of Unities. At least in respect of these important difficulties, the theory of Unities shows itself therefore to be better suited to account for the way in which the existence of a whole depends on its parts fulfilling certain conditions (i.e. having certain properties and relations to one another), for the way in which different wholes can be superposed with one another, and for the way in which wholes are capable of surviving the loss and gain of parts.

With regard to conditioned sums the point was made that taking the whole to be a sum forces upon the theorist the consequence that the conditions (defined with reference to the parts) under which the whole is assumed to exist are extraordinarily, and implausibly, sensitive. The juxtaposition of a new molecule to some of the old ones, at the periphery of a human organism, was sufficient to disrupt the collective capacity of the countless trillions of old molecules to have a sum, or at least to have a human organism as a sum. The change in the condition of the old molecules must be viewed as insignificantly small (the change can perhaps be described as follows: Some of the old molecules at the periphery of the organism now have the relational property *is juxtaposed to a molecule of such and such kind in such and such direction*, in virtue of their relation with the new molecule, which they did not have before). The change in their collective capacity to make up something is great. The predicament is obviously difficult to accept.

The theory of Unities, by contrast, is consistent with assumptions under which this difficulty does not arise. If for example the old molecules are assumed to underlie the organism before the new molecule is added, this does not entail that after the addition of the new molecule the old molecules cannot continue to underlie the organism, even if it is assumed that the old molecules and the new one *taken together* now underlie the organism. Of course, it is admitted that the organism cannot be at the same time a sum of the former and a sum of the latter. However, wholes are not fundamentally sums, and so the organism may be made up of the old molecules (by being underlain by them) without being a sum of those molecules.

Maintaining the notion of a non-unique sum, it was argued, implies the claim that distinct entities may be presently indiscernible, a claim that appears very dubious on closer examination. The theory of Unities, by contrast, makes ample provision for entities to be superposed with one another without being presently indiscernible. In section II of Chapter 9 it was pointed out that a Unity may be superposed with a distinct Unity, or with a collection. Whenever two entities u and v are superposed, it was assumed that for some xs both u and v are sums of the xs (see D17). Such superposition is fully consistent, however, with present discernibility. For u and v necessarily stand in different relations to the xs.

Consider for example the illustration in section III of Chapter 9 (Figure 1). A is a sum of $a_1 \ldots a_6$, and B is a sum of $a_1 \ldots a_6$. However, each of the entities, $a_1 \ldots a_6$ bears to A, and *not* to B, the relation *is an element of*. Each of the entities $a_1 \ldots a_6$ bear to B only the relation *is a pre-element of*. Moreover, A and B are further discernible in that b_1 and b_2 are elements of B but not of A. These are differences between A and B which do not concern properties or relations they have at times other than the present, or in worlds other than the actual one.

Finally, let us compare neoclassical mereology and the theory of Unities as regards mereological variation. It was shown in section III of Chapter 7 that the neoclassical notion of a mereologically varying sum runs into difficulties in connection with the process whereby a part is lost (similar problems arise in connection with the converse process). It transpired that if it is accepted that the whole must be *determinately* a whole throughout the process, then there are three options: The whole must either cease to exist during a transitional period, or it must be made up both of the old parts and of the old parts + new parts, or it must flip *instantaneously* from being made up of the old parts to being made up of the old parts + new parts.

Under neoclassical assumptions, none of these options seemed tenable. Under Unity-theoretical assumptions, by contrast, one immediately notices that the second of the three options is acceptable. Suppose a car is conceived of as a Unity, and suppose it has five elements: The body and the four wheels . This implies that the car is made up of (interpreting 'made up of' as 'underlain by') the body and the four wheels. It is consistent, however, with the assumption that the car is at the same time made up of the body and *three* of the wheels. The fact that being underlain by the body and four wheels is compatible with being underlain at the same time by the body and three wheels, allows the Unity-theorist to avoid the implausibilities of intermittent existence or instantaneous change, not to speak of outright contradiction, which are the main options of the neoclassical theorist.

III - The Paradox of Tibbles

Noteworthy consequences of the switch to a conception of wholes as Unities as regards fundamental cruxes of the metaphysics of concrete entities are revealed by considering that famous creature of thought-experiment, Tibbles the cat. The case of Tibbles is a classical philosophical paradox, where reflection on the simplest of our everyday beliefs leads to a striking contradiction, the avoidance of which does not seem possible without departing more or less radically from common sense-views. It has therefore been possible, by means of an analysis of this or closely analogous examples, to argue for a surprising range of *prima facie* implausible positions, such as the four-dimensionalist view of concrete entities, the sortal relativity of identity, mereological essentialism, or the view that concrete entities may coincide with one another.[15] I shall examine the paradox of Tibbles (which I shall refer to simply as 'the Paradox') in what follows, indicating the way it can be resolved under the assumptions of the theory of Unities, comparing this with resolutions based on the conception of wholes as sums.

Tibbles is a cat sitting on a mat at time *t*. Later, at *t*', it is again sitting on the mat. Only in the mean time it has lost its tail. The Paradox is derived by assuming that an entity called 'Tib' exists, which comprises all of Tibbles' parts at *t*', and pointing out that on the evidence available at *t* Tibbles and Tib are distinct from one another, while on the evidence

[15] Mark Heller (1988) argues from this paradox that concrete entities are not continuants, but rather have temporal parts; Peter Geach uses a variant of the example to argue for the sortal relativity of identity (see Geach 1980, 215f.); Roderick Chisholm uses it to support his mereological essentialist view (see Chisholm 1979); Peter Simons, and E.J. Lowe, argue for the coincidence of concrete entities on the basis of the Tibbles example (see Simons 1987, 117-127; Lowe 1989, Chapter 6).

available at *t'* they are identical to one another, thereby arriving at a contradiction.

Suppose we consider Tibbles (uncontroversially) to be composed of the following parts at *t*: Head, Neck, Torso, Leg1, Leg2, Leg3, Leg4, and Tail.[16] The foregoing discussion allows us to distinguish between two different things that 'Tib' might be taken to refer to in the context of the example. On the one hand (a) it might be taken to refer to *a whole composed of* Head, Neck, Torso, Leg1, Leg2, Leg3, Leg4. On the other hand (b) it might be taken to refer to *a whole which corresponds to* (or, equivalently, to *a whole which is made up of*) Head, Neck, Torso, Leg1, Leg2, Leg3, Leg4.

Usually this distinction is not pointed out, presumably because on the prevalent conception of wholes as sums the two descriptions are equivalent. On Unity-theoretical assumptions, however, a whole which corresponds to the *x*s is not necessarily a whole which is composed by the *x*s (recall that we deny assumption D4 of section II, Chapter 9). Therefore, according to the theory of Unities the descriptions in (a) and (b) are not equivalent.

To present the Paradox with these comments in mind in terms that are not question begging, let us use four names, 'Tibbles', 'Tib', 'Tibbles*' and 'Tib*', explaining their references by means of associated descriptions (note that alternative readings of 'Tib' and 'Tib*' are provided). To facilitate the formulation of such descriptions, let us say that parts of a cat which do not overlap any part of a tail are 'bodily parts' of the cat:

Tibbles the cat sitting on the mat at *t*

Tib either (a) the whole composed the bodily parts of Tibbles
or (b) the whole corresponding to the bodily parts of Tibbles

Tibbles* the cat sitting on the mat at *t'*

[16] For the purposes of this explanation, viewing Tibbles as composed of any mutually disjoint parts would do just as well, provided there are at least two parts which are disjoint from Tail.

Tib* either (a) the whole composed of the bodily parts of Tibbles*
or (b) the whole which corresponds to the bodily parts of Tibbles*

Following Simons, we may express the Paradox by means of the following *prima facie* compelling argument:[17]

According to the conception of wholes as sums, the two readings of 'Tib' and 'Tib*' (respectively) are equivalent. Therefore one assumes that Tibbles has parts which Tib does not have, and that they have different shapes, weights, overall lengths, etc. If that is so, it follows that

(1) Tibbles is not identical to Tib

Since Tibbles* and Tib* seem not to be discernible in any respect, one assumes that

(2) Tibbles* is identical to Tib*

Assuming Tibbles and Tib are continuants, we have good reason to believe (in the context of the fictional example) that

(3) Tibbles is identical to Tibbles*

and

(4) Tib is identical to Tib*

Finally, since Tibbles is identical (by (3)) to Tibbles*, which is identical (by (2)) to Tib*, which is identical (by (4)) to Tib, assuming the transitivity of identity we conclude that

(5) Tibbles is identical to Tib

And (5) obviously contradicts (1).

---

[17] See Simons 1987, 119.

As Simons points out, conclusion (5) might be blocked by denying any one of several intuitively appealing theses: By denying that concrete macroscopic entities exist; by denying that they are continuants; by denying that they may survive the gain or loss of parts; by denying the absoluteness of identity; or by denying the transitivity of identity.[18] I shall not discuss these responses here. The theory of Unities allows us to avoid each of these denials, and I would like to concentrate on a comparison with other responses which have this feature in common with the theory of Unities.

If one accepts the conception of wholes as sums, and yet wishes to avoid all of the denials mentioned in the former paragraph, one seems to have two basic options. The first is to claim that arbitrarily prescribed parts of concrete entities, Tib being an example, cannot be assumed to exist. This allows one to deny (4). This is van Inwagen's position, at least as expressed in 'The Doctrine of Arbitrary Undetached Parts'.[19] Let us call this option the 'No-Tib' view.

The alternative option is to claim that distinct concrete entities may coincide, or at least may be in the same place at the same time. This allows one to claim that Tibbles* coincides with Tib* even though they are not identical to one another, in other words, to deny (2). This is the position of Wiggins, Doepke, Simons and Lowe, among others. Let us call this option the 'Many-Tibbles' view.

Before examining the two views individually, it should be emphasised that because they both embrace the conception of wholes as sums, they both face the substantial difficulties confronting that conception especially with regard to mereological variation (and in the case of the Many-Tibbles view, also with regard to non-uniqueness of sums), difficulties which were discussed in Chapter 7. As we will see, however, the accounts offered of the Paradox according to these views lead (respectively) to additional problems and implausibilities.

1. The No-Tib View

According to the No-Tib view, no entity exists which we would refer to by means of 'Tib', a term whose denotation we attempt to specify as noted above by one of the two descriptions (a) or (b). However, intuitively

[18] Regarding the last four responses, see references in Simons 1987, 120. Regarding the first, see, e.g. Unger 1979 and 1981.

[19] Van Inwagen 1981.

speaking, we do not seem to have any difficulty in conceiving or imagining an entity thus described. In fact, on the basis of either description, we tend to imagine an entity which is very similar in many respects to Tibbles. It weighs almost as much as Tibbles; it is almost as long; it has four legs, a head, an abdomen, just like Tibbles; it has fur of precisely the same colour and texture. Indeed, on the face of it, we imagine Tib as being in every way similar to a tailless cat, except for the fact that it is itself a proper part of a cat, whereas tailless cats cannot be proper parts of cats any more than cats with tails can be.

The intuition supporting such a conception of Tib seems to be closely associated with the widespread use of sortal terms which apparently apply to entities that are (typically) undetached parts of other entities, such as 'hand', 'head', 'toe', 'branch', 'handle', 'blade'. It is true that such part-terms, as we may call them, are usually associated with relatively small parts of the respective wholes (a head, for example, is usually a relatively small part of the animal to which it belongs). But this is not always the case. The sword's blade may well be almost as large as the sword itself. Or, to take another example, the part of a sprawling building which we designate as the 'main building' may well be far larger than all the subsidiary wings of the building taken together.

Thus far I have been speaking intuitively. An adequate explanation of why we have this intuition (which we may call 'the Tib-intuition') would seem to be required of any acceptable account of the Paradox. If it is denied that we have this intuition, even an apparent paradox cannot be presented. If Tib were conceived from the start, *prima facie*, as being of a completely different character to Tibbles, then we would not even be *inclined* to assent to (4). The No-Tib theorist, however, is not denying that the Paradox seems paradoxical.

The No-Tib view does not deny, then, that we have the Tib-intuition, but it rejects it as being mistaken. Moreover, it rejects it radically. Not only were we not referring to a tailless cat, when we used 'Tib'. We were not referring to an entity of any sort. The No-Tib view is therefore strikingly counter intuitive in this respect.

It may well be that any solution to the Paradox is bound to be counter intuitive in one respect or another. However, it is a particular drawback of a solution if while denying a powerful intuition it does not provide a satisfactory explanation as to why we might have such an intuition. If we are not shown how it comes about that a certain intuition of ours commits a mistake, we are likely to continue to follow that intuition. The No-Tib view does not provide any such explanation, and it fails to assuage the intuition that it denies.

It might be suggested that we have the intuition because we use a term which makes apparent reference to a single entity. Perhaps a coherent account might be given of claims that we make apparently about a single entity, by showing how such claims can be translated (equivalently) into claims about many entities. Simons claims that van Inwagen's suggestion that such an account is available has not been demonstrated. In his later book, *Material Beings*, however, van Inwagen does work out such an account in some detail, an account which I shall not attempt to assess here.[20]

However, granting that such an account is available, it does not explain why such use of a term which makes only apparent reference to a single entity should engender, or be associated with, the (misguided) intuition that one is indeed referring to a single entity. The term 'flock' would seem to be an example of a term making apparent reference to a single entity (it is, like 'Tib', grammatically singular). However, it can hardly be claimed that use of this term is associated with the compelling intuition that a flock is a single entity.[21]

Apart from its sharp conflict with the Tib-intuition, and apart from the problems it shares with any view based on the conception of wholes as sums with regard to accounting for mereological change, the No-Tib view gives rise to several other problems.

First we note that the view involves the assumption that there are *some* undetached parts. Plausible candidates for undetached parts of Tibbles which *do* exist, according to van Inwagen, are whole molecules, whole cells, and whole organs (such as the heart). Such parts of course themselves have parts, in turn, parts which are (by the transitivity of *is a part of*) also parts of Tibbles.

Let us grant the No-Tib view that Tib does not exist, but let us assume that more or less straightforwardly separable (in the surgical sense) organs of Tibbles do exist. Thus suppose the No-Tib theorist accepts the existence of entities such as Tibbles' heart, liver, pancreas, and brain. Now, suppose further that Tibbles' brain, which we may call 'Brain', may

[20] See Simons 1987, 121 (n.40); van Inwagen 1990.

[21] Whether a flock is, indeed, a single entity, is a different question (I have discussed issues associated with this question in Part One). It is quite possible that even though we do not have a compelling intuition that a flock is a single entity, it is, as a matter of fact, a single entity.

survive in separation from the rest of Tibbles' body, if appropriately sustained by artificial means, perhaps in some futuristic vat. Finally, suppose that in such circumstances - where we might imagine the rest of Tibbles' body to be destroyed - Tibbles survives. We may now reformulate the Paradox, replacing 'Tib' with 'Brain' and 'Tib*' with 'Brain*'. In this case, *ex hypothesi*, the No-Tib theorist accepts the existence of Brain. So stage (4) of the Paradox argument cannot be denied, and contradiction results.

Of course the No-Tib theorist might deny also that organs such as Brain exist (though the more types of parts of Tibbles that are ruled out, the more difficult it would be to justify that Tibbles itself exist, for one would have to point out pertinent differences between those parts and Tibbles itself, which would justify the claim that Tibbles exists but not those parts). However, there seems to be no reason in principle why with respect to some types of parts of some entities a situation of the sort involving Tibbles and Brain should not obtain. And if it can obtain, in principle, then cases analogous to the Paradox can in principle occur, in connection with which the No-Tib view cannot avoid contradiction. This point alone would seem to be enough to show that the No-Tib view is untenable as a response to the Paradox.

The No-Tib view involves a further difficulty, however. There is much appeal in the suggestion that, crudely put, the existence of a whole can only be affected by what happens inside it, and not by what happens outside it (unless of course what happens outside it affects what happens inside it). To express this suggestion more precisely, let us use expressions of the form 'a change in the *x*s' as short for 'a change in the properties of, or relations between, the *x*s'. The following principle may then be formulated, which I call the 'Autarchy of Wholes':

<u>Autarchy of Wholes (I)</u>
for any *x*s, for any *y*s, for all *t*,
if there is a whole which corresponds to the *x*s at *t*, and if the *x*s are among the *y*s, then
a change (commencing at *t*) in the *y*s does not cause there to be no whole which corresponds to the *x*s, unless the change in the *y*s either causes or partially consists in a change in the *x*s

To illustrate the principle, suppose *a* is a table, a whole which corresponds to pieces of wood, the *b*s. Suppose *c* is a thin wooden board which has precisely the same surface shape and size as the table top. We do not think that by gluing *c* to *a* we have brought it about that there is no whole which corresponds to the *b*s. Indeed, we might suppose that that which is now (after the gluing) a whole which corresponds to the *b*s is not a table. We might claim, for instance, that it is a proper part of a table, and that a proper part of a table is not itself a table. And we might claim that after the gluing, *a* is no longer a whole which corresponds to the *b*s, but rather a whole which corresponds to the *b*s and *c*. These claims, however, are all compatible with there being still some whole which corresponds to the *b*s after the gluing.

The No-Tib view violates Autarchy of Wholes (I). It is easiest to see this if we consider the process involved in the Paradox of Tibbles in reverse. Suppose Tibbles has no tail, having lost it a short time ago. The tail is now surgically restored to Tibbles. Taking the conditions to be those of an ideal thought-experiment, we assume that the process of re-attaching the tail neither involves nor causes any change in the parts of Tibbles (none of which overlap a tail), the *p*s. Suppose the *q*s are the parts of Tibbles and the tail. Before the operation, there is a whole which corresponds to the *p*s, namely Tibbles. After the operation the No-Tib view takes it that there is no whole which corresponds to the *p*s. So there were changes in the *q*s, which caused there to be no whole which corresponds to the *p*s, and yet there were no changes in the *p*s - in violation of Autarchy of Wholes (I). Therefore, if one embraces that principle, then one has a further reason for rejecting the No-Tib view.

2. The Many-Tibbles View

According to the Many-Tibbles view, Tibbles* is not identical to Tib*. The possibility of denying their identity to one another, notwithstanding the fact that they occupy the same region of space at the same time, is afforded by the neoclassical mereological denial of the uniqueness of sums. Tibbles* and Tib* are both taken to be sums of the same parts, and yet, since sums need not be unique, it may be assumed that these sums are distinct from one another. Two versions of the Many-Tibbles view seem possible. On the Many-Tibbles$_1$ view, Tib (and thus also Tib*, given assumption (4) of the Paradox) is an entity of a fundamentally different kind from Tibbles. Tibbles is a cat, an entity susceptible to flux of parts. Tib is not a cat - it is mereologically rigid or invariable.

E. J. Lowe, who is clear about insisting on the Many-Tibbles$_1$ view, distinguishes between mereologically varying entities, which he calls 'integrates', and mereologically unvarying (and indeed invariable) entities, which he calls 'summative objects'.[22] Summative objects fall into two basic types, for which Lowe uses the terms 'collective' and 'aggregate' respectively. Collectives are conceived as classical mereological sums of their parts. They may exist equally when their parts are closely packed together, as when their parts are scattered. The conditions of identity of aggregates, by contrast, involve the requirement that the parts of the entity in some sense adhere to or are in contact with one another. Lowe assumes that Tibbles is an integrate, while Tib is an aggregate.

According to the second version of the Many-Tibbles view, Many-Tibbles_2, both Tibbles and Tib are entities susceptible to mereological flux. Such a position is suggested by David Wiggins' discussion of the need to distinguish between Tibbles and something he designates by means of the expression 'Tib + Tail'.[23] The former is distinguished from the latter not as being susceptible to mereological flux *per se*, but as being capable of surviving, in particular, the destruction of Tail. Tib+Tail, by contrast, according to identity conditions assumed in classical mereology with regard to sums, exists if and only if both Tib and Tail exist. However, since Wiggins takes both entities to be continuants (indeed, they typically coincide throughout the life of Tibbles), it would seem that both are assumed to admit of mereological change, and thus Tib in particular is assumed to admit of this. Simons, in his discussion of the example, is explicit that both Tib and Tail are subject to metabolism.[24]

Both Many-Tibbles_1 and Many-Tibbles_2 are construed in keeping with the conception of wholes as sums, and as such are subject to the criticisms elaborated in Chapter 7. This applies not only to the account of mereological change, to which the No-Tib view was also committed, but also to the account of coincidence. According to both versions of Many-Tibbles, the two coincident entities at *t*' are indiscernible with respect to their present properties at *t*', and thus they are saddled with the serious difficulties associated with this suggestion. Since No-Tib is not committed to coincident sums, it has an advantage in this respect over the Many-Tibbles views.

As with No-Tib, the accounts offered by the Many-Tibbles views of the Paradox lead both views into further implausibilities. As regards doing

---

22 See Lowe 1989, 89.

23 Wiggins 1979.

24 See Simons 1987, 191.

justice to what I have called the 'Tib-intuition', both views fare better than No-Tib. Indeed, Many-Tibbles$_2$ respects that intuition unqualifiedly. For Tib, on this view, is indeed similar to Tibbles in all the respects which the Tib-intuition takes it to be. Many-Tibbles$_1$, by contrast, can be seen as taking a middle position between the other two views. It accepts that there exists a single entity, to which we refer by means of the term 'Tib'. It rejects, however, that Tib is similar to Tibbles in respect of susceptibility to flux of parts.

The refusal of Many-Tibbles$_1$ to countenance flux of parts in Tib is not very happy, and it has an arbitrary flavour. Going along with the basic story of the Paradox, if after distinguishing between Tibbles and Tib we noticed that one of Tibbles whiskers has just been snipped off, we would be surprised at the suggestion that this resulted in the destruction of what we had taken to be Tib.

Moreover, the assumption that there is a category of concrete entities which are mereologically rigid, and yet which is, at the same time, emphatically distinguished from a classical mereological sum, seems very doubtful. After all, heaps of sand and lumps of clay - paradigmatic examples of Lowe's aggregates - *can* survive the loss of parts, at least if the every day usage of the terms 'heap' and 'lump' is anything to judge by. I do not destroy a sand dune if I walk across it and carry away a few grains which cling to my sandals, nor do I destroy a lump of clay by scraping a bit off with a knife. The assumption that there are such entities cannot provide a solid foundation for resolving the Paradox of Tibbles.

Furthermore, a principle only slightly less compelling than that of Autarchy of Wholes (I) is violated by Many-Tibbles$_1$. At *t*, there is a whole corresponding to the parts of Tib, which belongs to one ontological category (that of mereologically rigid entities). At *t'*, there is in addition a whole corresponding to the parts of Tib*, but this whole belongs to a category which is fundamentally different from that to which the other whole belongs. Again, a principle might be formulated as follows:

<u>Autarchy of Wholes (II)</u>
for any *x*s, for any *y*s, for all *t*,
if there is a whole which corresponds to the *x*s at *t*, and if the *x*s are among the *y*s, then
a change (commencing at *t*) in the *y*s does not cause there to be a whole of a fundamentally different type, which corresponds to the *x*s,

unless the change in the *ys* either causes or partially consists in a change in the *xs*

To the extent that we embrace this principle, we are thus faced with an additional reason for rejecting Many-Tibbles$_1$.

The second version of Many-Tibbles is free from this complaint. It faces a problem, however, which renders it if anything even more unpalatable than the first version. This can be seen by the simple means of envisioning a progressive series of losses of parts, each stage of which can be analysed in the manner presented in the Paradox.

Suppose the conditions at *t* are the same as those portrayed in the Paradox. At *t*', however, instead of losing its whole tail, we assume that Tibbles loses *half* the length of its tail. Thus at *t* we have according to the view we are considering two mereologically non-rigid entities, one of which has a complete tail, whereas the other is half-tailed: Tibbles and Tib1 (where 'Tib1' is explained in a way analogous to that of 'Tib' above). At *t*', both Tibbles and Tib1 are half tailed, but even though they coincide, they are not identical.

Now suppose that by *t*'' (a time later than *t*'), Tibbles has lost the remainder of its tail. In that case, on the same assumptions, there is some entity Tib2 (again, 'Tib2' is explained in a similar way), distinct from Tibbles, with which Tibbles coincides at *t*''. And what about Tib1? since we assume that it can survive the loss of parts, then we have no reason to deny that it too exists at *t*''. And yet, the reasoning of the Paradox (on the assumptions of Many-Tibbles$_2$) compels us to acknowledge that Tib1 is not identical to Tib2, even though it coincides with it. And since we already know from considering the first episode of the process that Tib1 is distinct from Tibbles, we conclude that at *t*'' we have at least three coincident entities, Tibbles, Tib1 and Tib2.

Since on this view the loss of a tail does not *generate* an additional cat (or cat-like entity, at least) but only reveals the necessity to recognise its existence, the conclusion that at *t*'' there exist at least three coincident entities follows irrespective of the process whereby the tail was lost (that is, whether or not it was actually lost gradually, in two stages).

A similar consideration, however, would show that the proliferation of coincident entities does not stop here. Far from it. For in a similar way we could conclude, considering Tibbles to lose its tail in the course of a hundred stages, that at the end of the process there are one hundred entities coincident with the cat. Indeed, assuming Tibbles has lost its tail, the number of coincident entities that have to be acknowledged is no smaller than the number of ways in which a part could be removed from the tail. A rough calculation shows that (assuming that parts no smaller than single atoms can be separated from one another) the number of entities which must be acknowledged as distinct from, but coincident with, the tailless Tibbles, is greater than the number of sand grains which would be needed to fill densely the volume of the known universe. It is hard to imagine that anyone could bring themselves to accept such a consequence in earnest.

In respect of ontological proliferation, then, we can distinguish between the three views, No-Tib, Many-Tibbles_1 and Many-Tibbles_2 as follows. Suppose Tibbles has lost its tail. According to No-Tib, there now exists Tibbles (as well as many proper parts of Tibbles - e.g. organs, cells, molecules, etc.). According to Many-Tibbles_1, there now exist two coincident entities Tibbles and Tib (as well as many proper parts of these). According to Many-Tibbles_2, there exist Tibbles, Tib1, Tib2, etc., i.e. innumerable coinciding entities. This is clearly a gradation of decreasing plausibility - other things being equal.

In respect of the adequacy of their account for the Tib-intuition, on the other hand, the views are related in an inverse gradation of implausibility. Many-Tibbles_2 justifies the intuition in full, Many-Tibbles_1 does so to a lesser degree, and No-Tib does so to the least degree.

3. The Unity-Theoretical View

Looking again at the alternative possible explanations of the use of the term ‘Tib’ in describing the Paradox, we find that if reading (b) is accepted, the theory of Unities offers us an immediate and simple resolution. Taking Tibbles to be a Unity, the theory allows us consistently to assume that Tibbles is at the same time both a whole which corresponds to the parts of Tibbles which include the parts of the tail, and a whole which corresponds to the parts of Tibbles which do not include the parts of the tail. In that case, we simply deny step (1) of the argument of the Paradox, for we affirm that Tib is identical to Tibbles.

Note that this provides us with a fully satisfying explanation of the Tib-intuition, according to which both Tibbles and Tib can survive mereological change, and both are, in fact, cats. Conceiving of Tibbles as a Unity, we may consider both Tib and Tibbles to be cats, indeed, we may consider them to be one and the same cat.

It is true that the Tib-intuition leads us to believe, *prima facie*, that Tib is a *tailless* cat, where as Tibbles has a Tail. So it leads us to believe, prima facie, that Tibbles is distinct from Tib. The Unity-theoretical view accounts for this *prima facie* belief by presenting the relations between Tibbles and Tib to be analogous to the relations between the Morning Star and the Evening Star.[25] In both cases, a *prima facie* belief that the entities involved are distinct from one another is grounded in the fact that they are referred to by means of different descriptions, descriptions which are compatible with, but do not imply, the assumption that the two entities are distinct from one another. The Morning Star is the star that appears in the morning, and the Evening Star is the star the appears in the evening. So they are *prima facie* distinct. But there is no contradiction in assuming that they are identical to one another.

Similarly, Tibbles is the whole which corresponds to parts which include parts of the tail, whereas Tib is the whole which corresponds to parts which do not include parts of the tail. Therefore, *prima facie* they are distinct from one another. However, the theory of Unities guarantees, by denying the assumption that wholes are fundamentally sums, that there is no contradiction in affirming nevertheless that Tibbles is identical to Tib.

In addition to fully satisfying the Tib-intuition (and in this respect being of comparable virtue to Many-Tibbles_2), it is remarkable that the Unity-theoretical view does not require the assumption that *any* coincident entities are involved in the story of Tibbles. Indeed, this view does not require us to acknowledge the existence of any entities other than those acknowledged by van Inwagen's minimalistic No-Tib view. In particular, we are relieved to avoid the radical implausibility of the consequence of Many-Tibbles_2, that there are innumerable mutually distinct entities which coincide with Tibbles after the loss of the tail.

But there is equally no need to presuppose the existence of doubtful mereologically rigid cat-like aggregates, which are required by Many-Tibbles_1. And even if we do believe that such aggregates exist, it is surely an advantage of the Unity-theoretical view that we are not compelled (at least on the grounds of considerations associated with the Paradox) to assume that they coincide with other, mereological varying entities. And furthermore, we are not committed to a violation of either version of the principle of Autarchy of Wholes, for we assume that one and the same

[25] See Frege 1892.

whole corresponds to the non-tail-parts of Tibbles, both before and after the loss of the tail.

We have noted in the foregoing discussion that the conception of wholes as Unities is compatible with the assumption that concrete entities may occupy the same place at the same time. It may well be, for example, that a statue is a Unity underlain by Torso, Head, Arm1, Arm2, etc., while the lump of clay that constitutes it is a different Unity underlain by clay particles, which are pre-elements, but not elements, of the statue. However, the Unity-theoretical solution to the Paradox allows us to maintain that the phenomenon of coincidence of concrete entities is not instantiated in typical cases involving mereological change. The phenomenon may be viewed as being much less widespread than it is according to views of such authors as Wiggins, Simons, and Lowe.

Finally, and importantly, as noted section II above, the conception of wholes as Unities is exempt from the more general weaknesses associated with the account of mereological change, superposition, and conditionality of sums, as offered by theories which embrace the framework of the conception of wholes as sums.

Now, the resolution of the Paradox in terms of the theory of Unities, along the lines just noted, depends on the assumption that explanation (b) of the use of 'Tib', in the context of the argument of the Paradox, is acceptable. Reading (b) does seem to me to be highly plausible, as a component of the explication of the terms used in formulating the Paradox. And we recall that on the conception of wholes as sums, reading (b) is indeed equivalent to reading (a). However, let us suppose that it is insisted that reading (a) rather than (b) is the one which the proponent of the Paradox has in mind. It is still the case, in my view, that the resources of the theory of Unities allow us to provide a more plausible treatment of the Paradox than the competing No-Tib and Many-Tibbles views.

If we assume reading (a), then Tib is supposed to be a whole which does not overlap the tail. The theory of Unities recognises such a whole - namely, a collection of proper parts of Tibbles, none of which overlap the tail. Indeed, there may be many such collections, which are superposed with one another (in the sense of 'superpose' explained in section III, Chapter 9). Furthermore, such collections, I have argued, are ontologically innocent, for any collection construed in accordance with the assumptions laid down above is identical to its members (taken collectively).

Collections, since they are taken to comply with classical mereological principles, are assumed to survive the dispersal of their members. In this respect they are comparable to collectives, rather than to aggregates (which do not survive the dispersal of their parts), in Lowe's theory. It might be thought that the assumption that Tib is a collection, or a collective, does less justice to the Tib-intuition than the assumption that it is rather a mereologically rigid lump, or an aggregate. If that is so, it must be noted that the existence of such aggregates is consistent with the theory of Unities, so that if necessary, they might be pressed into service. Still, the Unity-theoretical view would be at an advantage over Many-Tibbles$_1$, since it is not subject to the general criticisms which apply to accounts of mereological change, coincidence, etc. which are based on the conception of wholes as sums.

A resolution of the Paradox along the lines proposed, assuming explanation (a) of the use of 'Tib', is by no means unsatisfactory, when compared to the resolutions offered by the other views. However, the resolution offered on the assumption that the use of 'Tib' is rather explained by (b) is much more attractive in its parsimoniousness and simplicity. Although reading (b) may sound less natural, I think it faithfully expresses the observations which lead a common sense progression of thought into the Paradox of Tibbles. Bearing all this in mind, reading (a) together with the associated Unity-theoretical resolution of the Paradox have the most to recommend them among the positions which seem to be available.

Bibliography

Alexander, S. 1920, *Space, Time and Deity*, Vol. 2, London: Macmillan.

Allen, R.E. 1983, *Plato's Parmenides, Translation and Analysis*, Oxford: Basil Blackwell.

Aristotle, *Metaphysics*: Text - in Ross, W.D. 1924. *Aristotle's Metaphysics: A Revised Text with Introduction and Commentary* - Vols. 1 and 2, Oxford Classical Texts, Oxford: Clarendon press. Translation - by W.D. Ross, in Barnes 1984.

— *Poetics*: Text - in Bywater, I. 1909. *Aristotle on the Art of Poetry: A Revised Text with Critical Introduction, Translation and Commentary*, Oxford Classical Texts, Oxford: Clarendon Press. Translation - by I. Bywater, in Barnes 1984.

Armstrong, D. 1978, *Universals and Scientific Realism* - Vol. 1: *Nominalism and Realism*; Vol. 2: *A Theory of Universals*, Cambridge, Cambridge University Press.

— 1989, *Universals: An Opinionated Introduction*, Boulder: Westview Press.

— 1991, 'Classes are States of Affairs', *Mind* 100: 189-200.

— 1993. Response to Lewis in Bacon, Campbell *et al.* 1993.

Asch, S.E. 1968, 'Gestalt Theory', in D.L. Sills (ed.) *International Encyclopedia of the Social Sciences*, Vol. 6, The Macmillan Company & The Free Press.

Bacon, J., Campbell, K., and Reinhardt, L. 1993 (eds.), *Ontology, Causality and Mind: Essays in Honour of D.M. Armstrong*, Cambridge: Cambridge University Press.

Barnes, J. 1984 (ed.), *The Complete Works of Aristotle: The Revised Oxford Translation* - Vol. 2, Bollingen Series 71-2, Princeton: Princeton University Press.

Baxter, D. 1988, 'Identity in the Loose and Popular Sense', *Mind* 97: 575-82.

Beiser, F.C. 1993 (ed.), *The Cambridge Companion to Hegel*, Cambridge: Cambridge University Press.

Benziger, J. 1951, 'Organic Unity: Leibniz to Coleridge', *Publications of the Modern Language Association* 66: 24-48.

Bigelow, J. 1988, *The Reality of Numbers*, Oxford: Clarendon Press.

Black, M. 1971, 'The Elusiveness of Sets', *Review of Metaphysics* 24: 614-36.

Blyth, D.J. 1994, 'Wholes, Parts, and Sequences in Aristotle', *International Philosophical Quarterly* 136: 453-63.

Bogaard, P.A. 1979, 'Heaps of Wholes: Aristotle's Explanation of Compound Bodies', *Isis* 70: 11-29.

Bolzano, Bernard, 1851, *Paradoxes of the Infinite*, trans. by Fr. Prihonsky, intro. by D.A. Steele, 1950, London: Routledge and Kegan Paul.

Boolos, G. 1984, 'To Be is to Be a Value of a Variable (or to Be some Values of some Variables)', *Journal of Philosophy* 81: 430-49.

Bostock, D. 1988, *Plato's Theaetetus*, Oxford: Clarendon Press.

Broad, C.D. 1933, *An Examination of McTaggart's Philosophy* - Vol. 1. Cambridge: Cambridge University Press.

Burge, T. 1977, 'A Theory of Aggregates', *Noûs* 11: 97-117.

Burke, M.B. 1992, 'Copper Statues and Pieces of Copper: A Challenge to the Standard Account', *Analysis* 52: 12-17.

Burnyeat, M. 1990, *The Theaetetus of Plato* (with a translation by M.J. Levett, revised by M. Burnyeat), Indianapolis: Hacking Publishing Company.

Burton, David M.1985, *The History of Mathematics: An Introduction*, (1991 reprinting) Dubuque: Wm. C. Brown Publishers.

Cantor, G. 1895, *Contributions to the Founding of the Theory of Transfinite Sets*, trans. By P.E.B. Jordain, La Salle, Ill.: Open Court, 1952.

Chisholm, R. 1973, 'Parts as Essential to their Wholes', *Review of Metaphysics* 26: 581-603.

— 1976, *Person and Object: A Metaphysical Study*, London: Allen & Unwin.

— 1979. 'Objects and Persons: Revisions and Replies', in Sosa, E. 1979. (ed.) *Essays on the Philosophy of Roderick M. Chisholm*, Amsterdam: Rodopi.

The Concise Oxford Dictionary of Current English, 3rd edition, 1934, Oxford: Clarendon Press

Cornford, F.M, 1935. *Plato's Theory of Knowledge: The Theaetetus and the Sophist of Plato*, London: Routledge & Kegan Paul.

Crane, T. and Mellor, D.H. 1990, 'There is No Question of Physicalism', *Mind* 99: 185-206.

Cruse, D.A. 1979, 'On the Transitivity of the Part-Whole Relation', *Journal of Linguistics* 15: 29-38.

Dauben, J. W. 1979, *Georg Cantor: His Mathematics and Philosophy of the Infinite*, Cambridge Mass.: Harvard University Press.

Doepke, F. 1982, 'Spatially Coinciding Objects', *Ratio* 24, 45-60.

— 1986,'In Defence of Locke's Principle: A Reply to Peter M. Simons', *Mind* 95: 238-41.

— 1991, Review of Peter Simons's *Parts: A Study in Ontology*, *Noûs* 25: 393-96.

Eberle, R.A. 1970, *Nominalistic Systems*, Dordrecht, Reidel.

Ehrenfels, C. von, 1890, 'Über Gestaltqualitäten', *Vierteljahrsschrift für wissenschaftliche Philosophie* 14: 249-92. English translation - Smith, B. 1987, 'On Gestalt-Qualities', in Smith 1987.

Eklund, M. 2002, 'Peter van Inwagen on Material Beings', *Ratio* 15: 245-256.

Feldman, F. 1995, 'Life', in Kim and Sosa 1995.

Fine, G.J. 1979, 'Knowledge and *Logos* in the *Theaetetus*', *Philosophical Review* 88: 366-97.

Fine, K. 1995, 'Part and Whole', in Smith, B. and Woodruff, D. (eds.) *The Companion to Husserl*, Cambridge: Cambridge University Press.

Frege, G. 1884 . *Die Grundlagen der Arithmetik: Eine logisch-mathematische Untersuchung über den Begriff der Zahl*, Breslau: Koebner. English translation - Austin, J. L. 1953. (trans.) *The Foundations of Arithmetic*, 2nd edition, Oxford: Basil Blackwell.

— 1892, 'Über Sinn und Bedeutung', in *Zeitschrift für Philosophie und philosophische Kritik* 100: 25-50. English translation - Geach, P. and Black, M. 1960. 'On Sense and Reference', in Geach and Black 1960.

— 1895, 'Kritische Beleuchtung einiger Punkte in E. Schröder's *Vorlesungen über die Algebra der Logik*', *Archiv für Systematische Philosophie* 1: 433-56. English translation - Geach, P. and Black, M. 1960. 'A Critical Elucidation of some Points in E. Schroeder's Vorlesungen ueber die Algebra der Logik', in Geach and Black 1960.

Fuller, Andrew R. 1990, *Insight into Value: An Exploration of the Premises of a Phenomenological Psychology*, Albany: SUNY Press.

Furth, M. 1988, *Substance, Form, and Psyche: An Aristotelian Metaphysics,* Cambridge: Cambridge University Press.

Gallop, D. 1963, 'Plato and the Alphabet', *Philosophical Review* 72: 364 -76.

Geach, P. 1980, *Reference and Generality*, 3rd edition, Ithaca: Cornell University Press.

Geach, P., and Black, M. 1960, (trans.) *Translations from the Philosophical Writings of Gottlob Frege*, Oxford: Basil Blackwell.

Goodman, N. 1951, *The Structure of Appearance*, Cambridge, Mass.: Harvard University Press.

— 1956. 'A World of Individuals', in Bochenski, I.M., Church, A., and Goodman, N. *The Problem of Universals: A Symposium* (Aquinas Symposium, University of Notre Dame, 1956), Notre Dame, Ind.: University of Notre Dame Press. Reprinted in Benacerraf, P., and Putnam, H. 1964. (eds.) *Philosophy of Mathematics: Selected Readings*, Englewood Cliffs, N.J.: Prentice-Hall.

Grelling, K. and Oppenheim, P. 1938, 'Der Gestaltbegriff im Lichte der neuen Logik', *Erkenntnis* 7: 211-25. English translation - Simons, P. 1987. 'The Concept of Gestalt in the Light of Modern Logic', in Smith 1987.

— 1987. 'Logical Analysis of 'Gestalt' as 'Functional Whole'', in Smith 1987.

Griffin, N. 1977. *Relative Identity*, Oxford: Clarendon Press.

Grossmann, R. 1983, *The Categorial Structure of the World*, Bloomington: Indiana University Press.

Hale, B. 1987, *Abstract Objects*, Oxford: Basil Blackwell.

Halper, E. 1989, *One and Many in Aristotle's Metaphysics: The Central Books*, Columbus: Ohio State University Press.

Hamilton, E., and Cairns, H. 1961 (eds.), *The Collected Dialogues of Plato Including the Letters*, Bollingen Series 71, Princeton: Princeton University Press.

Hartshorn, C. 1942a, 'Organic and Inorganic Wholes', *Philosophy and Phenomenological Research* 3: 127-36.

— 1942b, 'The Group Mind', *Social Research* 9: 248-65.

Haslanger, S. 1994, 'Parts, Compounds, and Substantial Unity', in Scaltsas, Charles *et al.* 1994.

Healey, R.A. 1991, 'Holism and Nonseparability', *Journal of Philosophy* 88: 393-421.

Hegel, G.W.F. 1807, *Phenomenologie des Geistes*. English version - translated by A.V. Miller, with analysis and foreword by J.N. Findlay 1977, *Hegel's Phenomenology of Spirit*, Oxford: Clarendon Press.

Heller, M. 1988, *The Ontology of Physical Objects: Four-Dimensional Hunks of Matter*, Cambridge: Cambridge University Press.

— 1991. Review of Peter Simons's *Parts: A Study in Ontology*, *The Philosophical Review* 100: 488-90.

Henry, D. P. 1991, *Medieval mereology*, Amsterdam: B.R. Gruner.

Hughes, C. 1986, 'Is a Thing just the Sum of it Parts?', *Proceedings of the Aristotelian Society*, supplementary volume 85: 213-235.

Husserl, E. 1891, 'Besprechung von E. Schröder, *Vorlesungen über die Algebra der Logik (Exakte Logik)*, I. Band, Leipzig 1890', *Göttinger gelehrte Anzeigen*, 243-78. English translation - Willard, D. 1978 (transl.) 'Review of Ernst Schröder's *Vorlesungen über die Algebra der Logik*', *The Personalist* 59: 115-43.

— 1901, *Logische Untersuchungen: Untersuchungen zur Phänomenologie und Theorie der Erkenntnis*. 2nd edition - 1913. Panzer, U. 1984 (ed.). The Hague: Nijhoff. English translation - Findlay, J.N. 1970 (transl.). *Logical Investigations*, London: Routledge & Kegan Paul.

Johnson, W.E. 1921, *Logic* - Part I, Cambridge: Cambridge University Press.

Johnston, M. 1992, 'Constitution is not Identity', *Mind* 101: 89-105.

Kim, J. 1993, *Supervenience and Mind*, Cambridge: Cambridge University Press.

Kim, J. and Sosa, E. 1995 (eds.), *A Companion to Metaphysics*, Blackwell Companions to Philosophy, Oxford: Blackwell Reference.

Köhler, W. 1929, *Gestalt Psychology*, New York: H. Liveright.

Kripke, S. A. 1972, 'Naming and Necessity', in D. Davidson and G. Harman (eds.) *Semantics of Natural Language*, Dordrecht: Reidel. Reprinted in Kripke, S.A. 1980. *Naming and Necessity*, Oxford: Blackwell.

Küng, G. 1967, *Ontology and the Logistic Analysis of Language: An Enquiry into the Contemporary Views on Universals*, revised edition, Dordrecht: Reidel.

Leibniz, 1714, *Monadology*, in Parkinson, G.H.R. (ed. and transl.), and Morris, M. (transl.) *Leibniz: Philosophical Writings*, London: Dent.

Leonard, H.S., and Goodman, N. 1940, 'The Calculus of Individuals and its Uses', *Journal of Symbolic Logic* 5: 45-55.

Lerner, D. 1963(ed.), *Parts and Wholes*, The Colloquium on Scientific Method and Concept, New York: Free Press of Glencoe.

Leśniewski, S. 1916: 'Foundations of the General Theory of Sets. I', in *Collected Works*, Volume I, edited by Stanisław J. Surma (et al), 1992, Dordrecht: Kluwer Academic Publishers.

Lewis, D. 1986a, 'Against Structural Universals', *Australasian Journal of Philosophy* 64: 25-46.

— 1986b, *On the Plurality of Worlds*. London: Basil Blackwell.

— 1991, *Parts of Classes*, Oxford: Basil Blackwell.

— 1993. 'Many, but Almost One', in Bacon, Campbell *et al.* 1993.

Lloyd Morgan, C. 1923, *Emergent Evolution*, The Gifford Lectures Delivered in the University of St. Andrews in the Year 1922, London: Williams and Norgate.

Locke, J. 1689, *An Essay Concerning Human Understanding*. Edited with a Foreword by P.H. Nidditch, 1975, Oxford: Oxford University Press.

Lowe, E.J. 1982, 'The Paradox of the 1,001 Cats', *Analysis* 42: 27-31.

— 1989. *Kinds of Being: A Study of Individuation, Identity and the Logic of Sortal Terms*, Aristotelian Society Series, Vol. 10, Oxford: Basil Blackwell.

— 1995, 'Coinciding Objects: In Defense of the 'Standard Account', *Analysis* 55, 171-178.

Machover, M. 1996. *Set Theory, Logic and their Limitations*, Cambridge: Cambridge University Press.

McDowell, J. 1973, *Plato: Theaetetus*, translation with notes, Oxford: Clarendon Press.

McGrath, M., 1998, 'Van Inwagen's Critique of Universalism', *Analysis* 58: 116-121.

McTaggart, J.M.E. 1921, *The Nature of Existence* - Vol. 1, Cambridge: Cambridge University Press.

Meirav, A. 2000a, 'Non-unique Composition', *Synthese* 124: 323-342.

— 2000b, 'A Mereological Criterion for Physicality', *Southern Journal of Philosophy* 37: 619-631.

Mellor, D.H. 1995, *The Facts of Causation*, London: Routledge.

Miller, F.D. 1978, 'Aristotle's Use of Matter', *Paideia* 7: 105-119.

Miller, M. 1992, 'Unity and *Logos*: A Reading of *Theaetetus* 201c-210a', *Ancient Philosophy* 12: 87-111.

Moore, G. E. 1903a, *Principia Ethica*, Cambridge: Cambridge University Press.

— 1903b, 'The Refutation of Idealism', reprinted in Moore, G.E. 1922. *Philosophical Studies*, London: Routledge and Kegan Paul.

Morrow, G.R. 1970, 'Plato and the Mathematicians: An Interpretation of Socrates' Dream in the *Theaetetus*', *Philosophical Review* 79: 309-33.

Nagel, E. 1952, 'Wholes, Sums and Organic Unities', *Philosophical Studies* 3: 17-32. Reprinted in Lerner 1963.

Noonan, H. 1993, 'Constitution is Identity', *Mind* 102: 135-46.

Oelschlaeger, M. 1991, *The Idea of Modernism: From Prehistory to the Age of Ecology*, New Haven: Yale University Press.

Ojeda, A. E., 1993, *Linguistic Individuals*, Stanford: CSLI Publications.

Palmer, S. and Rock, I. 1990, 'The Legacy of Gestalt Psychology', *Scientific American* 263 (No. 6): 48-61.

Plato, *Parmenides*: Text - in Burnet, J. 1901. *Platonis Opera* - Vol. 2, London, Oxford University Press. Translation - Cornford, F.M., in Hamilton and Cairns 1961.

— *Theaetetus*: Text - in Burnet, J. 1901. *Platonis Opera* - Vol. 1, London, Oxford University Press. Translation - Levett, M.J. (revised by Burnyeat, M.), in Burnyeat 1990.

— *Sophist*: Text - in Burnet, J. 1901. *Platonis Opera* - Vol. 1, London: Oxford University Press. Translation - Cornford, F.M., in Hamilton and Cairns 1961.

Pollard, S. 1990: *Philosophical Introduction to Set Theory*, Notre Dame: University of Notre Dame Press.

Rescher, N. 1955, 'Axioms for the Part Relation', *Philosophical Studies* 6: 8-11.

— 1967. *The Philosophy of Leibniz*, Englewood Cliffs, N.J., Prentice Hall.

Rock, I. 1984, *Perception*, New York: Scientific American Library.

Rosenberg, J.F. 1995, 'Determinate/Determinable', in Kim and Sosa 1995.

Ross, W.D. 1924, *Aristotle's 'Metaphysics'*, 2 Volumes, Oxford: Clarendon Press.

Ruben, D. 1983, 'Social Wholes and Parts', *Mind* 92: 219-38.

Russell, B. 1903, *The Principles of Mathematics*, London: Allen & Unwin.

— 1919. *Introduction to Mathematical Philosophy*, London, Allen & Unwin.

Russell, B., and Whitehead, A.N. 1910-1913, *Principia Mathematica*, Cambridge, Cambridge University Press.

Ryle, G. 1960. 'Letters and Syllables in Plato', *Philosophical Review* 69: 431-51.

— 1990, 'Logical Atomism in Plato's *Theaetetus*', *Phronesis* 35: 21-46.

Sanford, D. H. 1993, 'The Problem of the Many, Many Composition Questions, and Naive Mereology', *Noûs* 27: 219-228.

Sayre, K.M. 1969, *Plato's Analytic Method*, Chicago: University of Chicago Press.

Scaltsas, T. 1990, 'Is a Whole Identical to its Parts?', *Mind* 99: 583-98.

— 1992, 'Substratum, Subject, and Substance', in Preus, A. and Anton, J. (eds.). 1992. *Aristotle's Ontology*, Albany: State University of New York Press.

— 1994a. 'Substantial Holism', in T. Scaltsas, D. Charles *et al.* 1994.

— 1994b. *Substances and Universals in Aristotle's Metaphysics*, Ithaca: Cornell University Press.

Scaltsas, T., Charles, D., and Gill M.L. 1994. (eds.) *Unity, Identity, and Explanation in Aristotle's Metaphysics*, Oxford: Oxford University Press.

Schlick, M. 1935, 'Über den Begriff der Ganzheit', *Wissenschaftlicher Jahresbericht der Philosophischen* Gesellschaft an der Universität zu Wien - Ortsgruppe Wie der Kant-Gesellschaft für das Vereinsjahr 1933/4 und 1934/5: 23-37. Reprinted in Topitsch, E. (ed.). 1965, *Logik der Sozialwissenschaften*, Köln, Berlin: Kiepenheuer & Witsch. English translation - 'On the Concept of Wholeness' in *Philosophical Papers* - Vol. 2, edited by Mulder, H.L. and van de Velde-Schlick, B.F.B. 1979, Dordrecht: Reidel.

Schröder, E. 1890-1910. Vorlesungen über die Algebra der Logik (Exakte Logik), Leipzig: Teubner.

Shusterman, R. 1988, 'Nietzsche and Nehamas on Organic Unity', *Southern Journal of Philosophy* 26: 378-92.

Simons, P. 1982, 'The Formalisation of Husserl's Theory of Wholes and Parts', in Smith 1982.

— 1987, *Parts: A Study in Ontology*, Oxford: Oxford University Press.

— 1991. Review of David Lewis's *Parts of Classes*, *Mind* 100: 394-97.

Sinisi, V.F. 1983a, 'Leśniewski's Foundations of Mathematics', *Topoi* 2: 3-6.

— 1983b, 'The Development of Ontology', *Topoi* 2: 53-61.

Smith, B. 1982 (ed.), *Parts and Moments: Studies in Logic and Formal Ontology*, Munich, Philosophia.

— 1988 (ed.), *Foundations of Gestalt Theory*, Munich: Philosophia.

— 1994, *Austrian Philosophy: The Legacy of Franz Brentano*, Chicago: Open Court.

— 1997, 'On Substances, Accidents and Universals: In Defense of a Constituent Ontology', *Philosophical Papers* 26: 105-127.
Sprigge, T.L.S. 1970, *Facts, Words and Beliefs*, London: Routledge.
Stenius, E. 1974, 'Sets', *Synthese* 27: 162-188.
Tiles, M. 1989, *The philosophy of set theory : an historical introduction to Cantor's paradise* , Oxford: Blackwell.
Unger, P. 1979, 'I Do Not Exist', in G.F. MacDonald *Perception and Identity: Essays Presented to A.J.Ayer with his Replies to them*, Ithaca: Cornell University Press.
— 1981. 'The Problem of the Many', in P.A. French, T.E. Uehling and H.K. Wettstein (eds.) *Midwest Studies in Philosophy*, vol. 6, Minneapolis: University of Minnesota Press.
Van Fraassen, B. 1966, 'Singular Terms, Truth-Value Gaps and Free Logic', *Journal of Philosophy* 63: 481-95.
Van Inwagen, P. 1981. 'The Doctrine of Arbitrary Undetached Parts', *Pacific Philosophical Quarterly* 62: 123-37.
— 1987. 'When are Objects Parts?', in Tomberlin, J.E. 1987 (ed.) *Metaphysics*, Philosophical Perspectives 1, Alascadero, Calif.: Ridgeview Publishing Company.
— 1990. *Material Beings*, Ithaca, N.Y.: Cornell University Press.
— 1993, 'Naive Mereology, Admissible Valuations, and Other Matters', *Noûs* 27: 229-234.
— 1994: 'Composition as Identity', *Philosophical Perspectives 8: Logic and Language*: 207-220.
Westphal, K. 1993, 'The Basic Context and Structure of Hegel's Philosophy of Right', in Beiser 1993.
Wartenberg, T. 1993, 'Hegel's Idealism: The Logic of Conceptuality', in Beiser 1993.
Wertheimer, M. 1925, 'Gestalt Theory', *Social Research* 11: 78-99.
Williamson, T. 1990, Review of Peter Simons's *Parts: A Study in Ontology*, *Grazer Philosophische Studien* 38: 201-10.
— 1994. *Vagueness*, London: Routledge.
Wiggins, D. 1968, 'On Being in the Same Place at the Same Time' *Philosophical Review* 77: 90-5.
— 1980, *Sameness and Substance*, Oxford, Basil Blackwell.

Index

PHILOSOPHICAL STUDIES SERIES

52. Jesús Ezquerro and Jesús M. Larrazabal (eds.): *Cognition, Semantics and Philosophy*. Proceedings of the First International Colloquium on Cognitive Science. 1992 ISBN 0-7923-1538-3
53. O.H. Green: *The Emotions*. A Philosophical Theory. 1992 ISBN 0-7923-1549-9
54. Jeffrie G. Murphy: *Retribution Reconsidered.* More Essays in the Philosophy of Law. 1992 ISBN 0-7923-1815-3
55. Phillip Montague: *In the Interests of Others.* An Essay in Moral Philosophy. 1992 ISBN 0-7923-1856-0
56. Jacques-Paul Dubucs (ed.): *Philosophy of Probability.* 1993 ISBN 0-7923-2385-8
57. Gary S. Rosenkrantz: *Haecceity.* An Ontological Essay. 1993 ISBN 0-7923-2438-2
58. Charles Landesman: *The Eye and the Mind.* Reflections on Perception and the Problem of Knowledge. 1994 ISBN 0-7923-2586-9
59. Paul Weingartner (ed.): *Scientific and Religious Belief.* 1994 ISBN 0-7923-2595-8
60. Michaelis Michael and John O'Leary-Hawthorne (eds.): *Philosophy in Mind.* The Place of Philosophy in the Study of Mind. 1994 ISBN 0-7923-3143-5
61. William H. Shaw: *Moore on Right and Wrong*. The Normative Ethics of G.E. Moore. 1995 ISBN 0-7923-3223-7
62. T.A. Blackson: *Inquiry, Forms, and Substances*. A Study in Plato's Metaphysics and Epistemology. 1995 ISBN 0-7923-3275-X
63. Debra Nails: *Agora, Academy, and the Conduct of Philosophy.* 1995 ISBN 0-7923-3543-0
64. Warren Shibles: *Emotion in Aesthetics*. 1995 ISBN 0-7923-3618-6
65. John Biro and Petr Kotatko (eds.): *Frege: Sense and Reference One Hundred Years Later.* 1995 ISBN 0-7923-3795-6
66. Mary Gore Forrester: *Persons, Animals, and Fetuses.* An Essay in Practical Ethics. 1996 ISBN 0-7923-3918-5
67. K. Lehrer, B.J. Lum, B.A. Slichta and N.D. Smith (eds.): *Knowledge, Teaching and Wisdom.* 1996 ISBN 0-7923-3980-0
68. Herbert Granger: *Aristotle's Idea of the Soul.* 1996 ISBN 0-7923-4033-7
69. Andy Clark, Jesús Ezquerro and Jesús M. Larrazabal (eds.): *Philosophy and Cognitive Science: Categories, Consciousness, and Reasoning*. Proceedings of the Second International Colloquium on Cogitive Science. 1996 ISBN 0-7923-4068-X
70. J. Mendola: *Human Thought.* 1997 ISBN 0-7923-4401-4
71. J. Wright: *Realism and Explanatory Priority*. 1997 ISBN 0-7923-4484-7
72. X. Arrazola, K. Korta and F.J. Pelletier (eds.): *Discourse, Interaction and Communication.* Proceedings of the Fourth International Colloquium on Cognitive Science. 1998 ISBN 0-7923-4952-0
73. E. Morscher, O. Neumaier and P. Simons (eds.): *Applied Ethics in a Troubled World.* 1998 ISBN 0-7923-4965-2
74. R.O. Savage: *Real Alternatives, Leibniz's Metaphysics of Choice*. 1998 ISBN 0-7923-5057-X
75. Q. Gibson: *The Existence Principle.* 1998 ISBN 0-7923-5188-6
76. F. Orilia and W.J. Rapaport (eds.): *Thought, Language, and Ontology.* 1998 ISBN 0-7923-5197-5

PHILOSOPHICAL STUDIES SERIES

77. J. Bransen and S.E. Cuypers (eds.): *Human Action, Deliberation and Causation.* 1998 ISBN 0-7923-5204-1
78. R.D. Gallie: *Thomas Reid: Ethics, Aesthetics and the Anatomy of the Self.* 1998 ISBN 0-7923-5241-6
79. K. Korta, E. Sosa and X. Arrazola (eds.): *Cognition, Agency and Rationality.* Proceedings of the Fifth International Colloquium on Cognitive Science. 1999 ISBN 0-7923-5973-9
80. M. Paul: *Success in Referential Communication.* 1999 ISBN 0-7923-5974-7
81. E. Fischer: *Linguistic Creativity.* Exercises in 'Philosophical Therapy'. 2000 ISBN 0-7923-6124-5
82. R. Tuomela: *Cooperation.* A Philosophical Study. 2000 ISBN 0-7923-6201-2
83. P. Engel (ed.): *Believing and Accepting.* 2000 ISBN 0-7923-6238-1
84. W.L. Craig: *Time and the Metaphysics of Relativity.* 2000 ISBN 0-7923-6668-9
85. D.A. Habibi: *John Stuart Mill and the Ethic of Human Growth.* 2001 ISBN 0-7923-6854-1
86. M. Slors: *The Diachronic Mind.* An Essay on Personal Identity, Psychological Continuity and the Mind-Body Problem. 2001 ISBN 0-7923-6978-5
87. L.N. Oaklander (ed.): *The Importance of Time.* Proceedings of the Philosophy of Time Society, 1995–2000. 2001 ISBN 1-4020-0062-6
88. M. Watkins: *Rediscovering Colors.* A Study in Pollyanna Realism. 2002 ISBN 1-4020-0737-X
89. W.F. Vallicella: *A Paradigm Theory of Existence.* Onto–Theology Vindicated. 2002 ISBN 1-4020-0887-2
90. M. Hulswit: *From Cause to Causation.* A Peircean Perspective. 2002 ISBN 1-4020-0976-3; Pb 1-4020-0977-1
91. D. Jacquette (ed.): *Philosophy, Psychology, and Psychologism.* Critical and Historical Readings on the Psychological Turn in Philosophy. 2003 ISBN 1-4020-1337-X
92. G. Preyer, G. Peter and M. Ulkan (eds.): *Concepts of Meaning.* Framing an Integrated Theory of Linguistic Behavior. 2003 ISBN 1-4020-1329-9
93. W. de Muijnck: *Dependencies, Connections, and Other Relations.* A Theory of Mental Causation. 2003 ISBN 1-4020-1391-4
94. N. Milkov: *A Hundred Years of English Philosophy.* 2003 ISBN 1-4020-1432-5
95. E.J. Olsson (ed.): *The Epistomology of Keith Lehrer.* 2003 ISBN 1-4020-1605-0
96. D.S. Clarke: *Sign Levels.* Language and Its Evolutionary Antecedents. 2003 ISBN 1-4020-1650-6
97. A. Meirav: *Wholes, Sums and Unities.* 2003 ISBN 1-4020-1660-3
98. C.H. Conn: *Locke on Esence and Identity.* 2003 ISBN 1-4020-1670-0

KLUWER ACADEMIC PUBLISHERS – DORDRECHT / BOSTON / LONDON

MIX
Papier aus verantwortungsvollen Quellen
Paper from responsible sources
FSC® C105338

If you have any concerns about our products,
you can contact us on
ProductSafety@springernature.com

In case Publisher is established outside the EU,
the EU authorized representative is:
Springer Nature Customer Service Center GmbH
Europaplatz 3, 69115 Heidelberg, Germany

Printed by Libri Plureos GmbH
in Hamburg, Germany